Growing Old in the future

Translation from Dutch by: C.N. ter Heide-Lopey

C.F. Hollander

H.A. Becker (Eds.)

Growing Old in the Future

Scenarios on health and ageing 1984-2000

Scenario-report, commissioned by the Steering Committee on Future Health Scenarios

1987
Martinus Nijhoff Publishers
A member of the Kluwer Academic Publishers Group
Dordrecht – Boston – Lancaster

for the United States and Canada: Kluwer Academic Publishers, P.O. Box 358, Accord Station, Hingham, MA 02018-0358, USA
for the UK and Ireland: Kluwer Academic Publishers, MTP Press Limited, Falcon House, Queen Square, Lancaster LA1 1RN, UK
for all other countries: Kluwer Academic Publishers Group, Distribution Center, P.O. Box 322, 3300 AH Dordrecht, The Netherlands

ISBN 0 89838 869 4

Library of Congress Catalog Card Number 87-11047

Martinus Nijhoff Publishers, P.O. Box 163, 3300 AD Dordrecht, The Netherlands.

PRINTED IN THE NETHERLANDS

Preface

This publication represents the report of the Scenario Committee on Ageing. The draft report was discussed with a wide range of experts, inter alia during the symposium 'Growing Old in the Future' held on October 27th, 1984. In addition to the scenario report, a background report containing the basic analysis employed in the scenarios on ageing has been prepared. The scenario report has been written in such a way that it can be read independently of the background study.

Scenarios are a relatively new phenomenon in health care and related policy. For this reason it might be useful to furnish the reader with a few suggestions.

As a first step, perusing the summary will provide the reader with an overall picture of this application of the scenario method in policy preparation and policy implementation in the sector health of the elderly.

As a second step, we would recommend that the scenario report be read in its totality. It might be useful to note down points on which the reader would like to make additions or variations.

Subsequently, it might prove beneficial to study both the scenario report and the background report in order to consolidate one's own ideas with respect to additions and variations, and to set down on paper in greater detail what might be the possible influences of contexts on patterns of care in order to be able to weigh the various effects. In this third stage, the as yet unpublished Policy Memorandum on Health in 2000 will no doubt also play a role. The reader will probably desire to study the policy memorandum in the light of the scenarios. In this stage, many readers will also want to take account of the reports of the other scenario committees. One thing that will be realized is that the four scenario reports only deal with a limited number of the aspects dealt with in the Policy Memorandum on Health in 2000. It is consequently the intention of the Steering Committee on Future Health scenarios also to devote scenarios studies to various other subjects.

During the implementation of the scenario project on ageing, information and co-operation was obtained from many sources. We would like to thank all concerned for their support and valuable contributions.

Scenario Commission on Ageing
Research Team

Leidschendam/Utrecht, January 1985

Table of contents

Page

Summary

1 Introduction

Steering Committee and Scenario Committees
Around the middle of 1983 the Steering Committee on Future Health Scenarios set up four scenario committees to intensify investigations of long-term developments in the health care sector. These committees were charged with drawing up scenarios with respect to (a) coronary and arterial diseases; (b) cancer; (c) life-styles; and (d) ageing.

Research questions
The scenarios on ageing aim at answering the following questions:
(1) What (future) developments will exert most influence on the health of the elderly in the Netherlands in the period 1984-2000?
(2) In view of the future health situation of the elderly and their increasing share in the Dutch population, what are the possible patterns of (health) care facilities in the period 1984-2000?

The first question relates to developments which to a large extent occur **autonomously** in present-day health care. The second question also serves to define the limits of this scenario project. The project does not include the choice of a certain pattern of (health) care facilities, or the choice of strategies for the realization of such a pattern. These choises will be dealt with in the Policy Memorandum on Health in 2000 of the Ministry of Health, Welfare and Culture.

Points of departure
The following points of departure were adhered to for the scenarios on ageing:

- The category of elderly was defined as people over fifty-five years of age. Where the use of (health) care facilities is discussed, in general the category concerned is 65 years of age and older.
- The time horizon chosen is the year 2000. Certain developments (demography, collective expenditure resulting from demographic developments) can be extrapolated till the year 2030.

- Where possible, future developments have been presented in quantitative terms, though it was often only possible to describe developments qualitatively.

The committee dealing with the scenario project has been careful to avoid generalizations on 'the question of the elderly'. By far the majority of elderly people are capable of living a healthy and independent life. Problems relating specifically to the elderly (illness, invalidity, social isolation, a lack of independence) only relate to some of them.

Definition of scenarios

A scenario is defined as 'a description of the present situation of the society (or part of it), of possible or desirable future situations of this society, as well as of series of events deriving from the present situation which could bring about those future situations'. The scenarios presented in this report focus on '**possible** future situations'. They are the most important possible 'contexts' whithin which (health) care facilities will be realized in the Netherlands till the year 2000. These 'contexts' as it were provide the actors in policy with respect to the elderly, including health care policy, with a 'learning environment'.
However, in the present scenarios no pronouncements are made as to the **desirability** of the future situations described. This, as well as the strategies for achieving the most desirable future situation, is left to the policy-makers. A separate report, The Background Study, gives "a description of the present situation of the society (or part of it)" for the subject ageing.
Figure I.1 in chapter 1 shows the clusters of variables which have been considered in this context.

Three context scenarios

After several rounds of designing, adapting and redesigning, the committee finally drew up three context scenarios, namely a reference scenario, a growth scenario and a shrinkage scenario. The terms growth and shrinkage relate to future demand for (health) care facilities. In this connection no account was taken of future supply of the facilities.
The scenarios show some degree of 'stylization' or 'over-accentuation'. This is typical of the scenario method.

2 Scenario A: The reference scenario

2.1 Introduction

The reference scenario is so designed that the other two scenarios can be contrasted with it. On the one hand it consists of a number of components which are assumed to remain constant, while on the other hand various existing trends are extrapolated into the future. Five aspects receive attention in the reference scenario, namely:

- demographic developments;
- the health of the elderly;
- social developments;
- demand by the elderly for facilities; and
- the economic context.

In this scenario it is assumed that medical-technological developments will remain constant till the year 2000.

2.2 Demographic developments

The medium variant of the 1980 population forecast of the Central Bureau of Statistics (CBS) has been used for estimating the future size and composition of the population.

Absolute and relative increase of the elderly
This forecast shows that there will be an increase both in the absolute number of elderly, and their relative share in the total population (ageing).
The category 'elderly persons' is characterized by an (absolutely and relatively) increasing number of **very old** people of 80 years of age and older (double ageing).
Furthermore, there will be a high surplus of women and an increasing number of persons living alone (single persons, widows/widowers and divorcees).

2.3 The state of health of the elderly

Indicators of the state of health

Four indicators have been chosen for describing the state of health of the elderly:

(a) Average life expectancy and maximum duration of life. A slight decrease in the difference in average life expectancy between men and women is foreseen as a result of the fact that the average life expectancy of men will rise from 72.8 to 73.3 years while among women it remains stable at 79.5 years. After 1990 - in accordance with the CBS forecast - no further increase in life expectancy is assumed. Man's maximum life-span is at present approximately 115 years. This also is expected to remain unchanged.

(b) Mortality according to causes. It is assumed that coronary and arterial diseases and cancer will continue to be the most important causes of death till 2000. As a consequence of ageing of the population there will be an absolute increase in the number of deaths per annum.

(c) Morbidity and invalidity. Characteristic of disease among the elderly is the relatively high incidence of chronic-degenerative diseases (calling for a high degree of aid) and a combination of diseases in one patient (multiple morbidity). Data relating to disease and invalidity among the elderly are inadequate. According to a recent study, the complaints most frequently encountered among elderly persons living independently are 'arthritis', 'shortness of breath', 'high blood pressure' and 'dizziness'. The relationship between complaints and invalidity is not self-evident. Disturbances in stamina, walking, functions of arm and hand (especially among women) and balance, are most often encountered. Four percent of elderly people of 55 and older who live independently require help in ADL. Among the elderly in homes this percentage is 53. Morbidity data relating to psychic disturbances (including dementia) are even scarcer than those relating to somatic complaints.
The percentages of elderly people who suffer from certain diseases or types of invalidity are assumed to remain constant in this scenario. As a consequence of the double ageing however, there is an absolute increase in the prevalence of sickness and invalidity among the population group 55 years of age and older.

(d) Subjective health. The manner in which a person perceives his own health is found to be not unequivocally related to age and sex. A person's perception of his state of health does not always coincide with his 'objective' state of health.

2.4 Social developments

In this section six social trends are outlined which are relevant not only for the reference scenario, but also for the growth and shrinkage scenarios (that is to say the direction of trends is the same for all scenarios, though the degree to which trends follow a certain direction may vary per scenario).

Level of education
There will be a heightening of the level of education of elderly people. As a result of this, people will behave 'younger' and participate more in adult education.

Emancipation of the elderly
The consequences of emancipation of the elderly are inter alia an increased degree of organization (in unions for the elderly), the emergence and training of leaders, a demand for representation in important councils (Social Economic Council, Sickness Fund Council), and the furthering of questions relating specifically to the elderly through existing political parties. More than this, the emancipation of the elderly will also have consequences on a cultural level.

Number of children
The decreased level of fertility will have consequences for the provision of umbrella care in as far as this must be provided by the children. In this connection a distinction is made between the various generations. Geographic mobility and the increasingly individualistic lifestyle exert a negative influence on possibilities for providing umbrella care. The percentage of elderly persons in the year 2000 who have (voluntarily) remained childless is estimated at 25%.

Emancipation of women
The process of emancipation of women will continue. Women are catching up on their educational lag, and despite increasing unemployment, there will be a further increase in the participation of (married) women in the labour market.
Emancipation of women will lessen the willingness and ability to provide informal aid.

Tolerance of euthanasia
The trend towards increasing tolerance of euthanasia will continue. Legislation governing the punishability of euthanasia will be relaxed. It will become acceptable for euthanasia to be quoted as a (contributory) cause of death. The increasing tolerance of euthanasia may exert an influence on the quality of life, or the ending of life, of the elderly. However, it is not to be expected that this attitude will influence the quantity of care of the elderly.

Upgrading of retirement and old age pensions
Expectations are that till the year 2000 there will be an increase in the number of people entitled to retirement pension, and that there will be an upgrading of these pensions. It is expected that measures will be speedily forthcoming to rectify the consequences for pension accrual of prolonged unemployment. Old age pensions will come under serious pressure. (As a consequence of demographic developments the ratio between the payers of premiums and those entitled to old age pensions will undergo an extremely negative influence.) However, in view of pressure from inter alia trade unions of the elderly, no significant deterioration of old age pensions is expected.

For purposes of the reference scenario other social developments (such as behaviour and attitudes with respect to health, participation in the labour force, pension age, solidarity within the generation of the elderly) are assumed to remain unchanged and are not discussed in this scenario.

2.5 The demand for (health) care facilities

The use made of (health) care facilities by the elderly increased sharply in the seventies. In recent years there has been a shift in this trend. With a view to furthering the independence of the elderly (but also to curbing government expenditure) an effort has been made to shift over from extramural to intramural facilities. Up to the present, however, there has been little progress in this direction, partly due to the fact that the demand for help has switched from informal care to extramural professional care, while the use of intramural facilities has hardly decreased.

Estimation methods for future demand for facilities
In this section developments in the demand for (health) care facilities by the elderly are outlined till the year 2000. For this purpose use has been made of two estimation methods:

- **The null variant:** application of constant age and sex-specific usage figures form a certain base year to the future population size and composition.
- **The trend variant:** extrapolation of past trends in age and sex-specific usage figures.

The use made at the present moment of (health) care facilities is the result of the demand for and the supply of these facilities. Results are compared with the policy recommended - the so-called policy variant - for indicating discrepancies between supply and demand in the future. In the reference scenario a selection of the facilities used by the elderly is made. Ten facilities are dealt with: three intramural, six extramural and one semimural.

GP care
Generally the GP is the first professional provider of aid who comes into contact with the health problems of the (older) population. The GP fulfills a number of functions in health care. The percentage of the population of 55 years of age and older who contact the GP in a given year varies from 68% (men of 55-59 years of age) to 85% (women from 75 years of age and older). There is no significant trend in the frequency of contact with GPs in recent years. Consequently the null variant coincides with the trend variant in the estimation of the future number of contacts with GPs from 14.5 million in 1982 to 17.9 million in 2000 for the age category 55 years of age and older. Ageing has consequences not only in a quantitative sense, but also in a qualitative sense: there is a shift in the morbidity pattern in the GP's practice towards chronic-degenerative complaints.
Pleas are heard from various quarters for more attention to geriatric problems in the training of GPs.
A few recent developments in health care are discussed in which GPs are involved. For instance there is greater co-operation within first echelon facilities (group practices, health centres). Furthermore, there are a number of experiments as to co-operation between GPs and second echelon care. The various forms of co-operation are stimulated by government (for instance through subsidizing).

Ambulatory mental health care
Extramural mental health care for the elderly takes place within the Regional Institutes for Ambulatory Mental Health Care (RIAGGs). At the present moment, 54 of the 59 recognized RIAGGs include separate units for psychogeriatry.

The goal of this type of aid is to prevent or at least to postpone admission of the elderly to intramural facilities. In 1982 sex and age-specific usage figures varied from 0.53% for men of 65-69 years of age, to 4.47% for women of 85 years of age and older.
Data on developments in the use of ambulatory mental health care facilities in the past are most inadequate. Consequently it is only possible to give an indication of future demand according to the null variant.
As a consequence of ageing of the population, in the period 1982-2000 there will be an increase from 23,000 to 41,000 in the number of elderly who make use of these facilities. A high degree of uncertainty however attaches to this estimate.
With respect to supply of psychogeriatric aid, the RIAGGs adhere to the so-called 20-20 norm (20 new patients per annum per 1000 elderly citizens and 20 hours aid per patient). At the moment the question of increasing the norm for the very elderly to 20-40 is under discussion. It is expected that in the future the socio-geriatric function of the RIAGGs will be intensified.

Homes for the elderly
Since 1977 the 7% norm has applied in homes for the elderly (possibilities must exist for caring for 7% of persons of 65 years of age and older in homes for the elderly). As of 1.1.1981 just over 133,000 persons of 65 years of age and older had been admitted (8.1%). In recent years the average age of patients admitted has risen (to slightly over 80), which has resulted in an increase in the need for nursing. The present capacity of homes for the elderly remains fairly constant at approximately 145,000 beds. This will necessitate adjustment of the system governing indications for admission. Table 1 shows the results of the projections.

Table 1 Estimation of the capacity of homes for the elderly (absolute and in percentages of the population of 65 years of age and older) in the period 1981-2000.

	null variant		trend variant		policy variant	
	abs.	%	abs.	%	abs.	%
1-1-1981	133,000	8.1	133,000	8.1	145,000	8.8
1-1-1985	151,000	8.7	136,000	7.9	145,000	8.4
1-1-1990	170,000	9.0	145,000	7.7	145,000	7.7
1-1-2000	196,000	9.4	(a)		145,000	6.9

a) see note 60 of chapter 2

In coming years the neighbourhood function of homes for the elderly will be stimulated. This concerns such functions as provision of meals, monitoring by means of alarm systems, short admission in case of illness, and socio-cultural activities involving elderly persons who live independently in the district.

Nursing homes
The total number of beds in nursing homes, approximately 47,500 (in 1983) is divided over nursing homes for somatic illnesses, nursing homes for psychogeriatric patients, and combined nursing homes. According to present norms for the number of beds permitted, the capacity for patients with somatic illnesses is too large, while that for psychogeriatric patients is too small. Estimates of total capacity according to the null and the trend variant are higher than permitted according to the policy variant.

Table 2 Estimate of capacity of nursing homes (somatic, psychogeriatric and combined) 1983-2000, absolute and in percentages of the population of 65 years and older

	null variant		trend variant		policy variant
	abs.	%	abs.	%	abs.
1-1-1983	47,456	2.80	47.456	2.80	46,471
1-1-1985	49,410	2.86	51,743	2.99	47,457
1-1-1990	55,169	2.91	59,053	3.12	51,669
1-1-2000	61,248	2.93	(a)		56,656

a) see note 60 of chapter 2

The role of the nursing home has come under discussion in recent years (costs aspect, privacy of patients). Partly on this account, nursing homes are introducing greater differentiation in the packet of services offered (day treatment, admission for the night, admission for the week-end, intervention in crises).

It is assumed that in the future differences between homes for the elderly and nursing homes will become less, a tendency which is stimulated by government policy. To achieve this, it will be necessary to finance the two types of institutions on a more equal footing.

Day treatment in nursing homes

The object of day treatment is to reactivate and resocialize patients whose condition would otherwise make them eligible for admission to a nursing home, in order to postpone permanent admission as long as possible.

As of 1.1.1984 there were 2225 day treatment places. According to the present policy norms, 2800 places are permitted for 1984, and nearly 3400 for the year 2000. Expansion is however dependent on the results of evaluation research.

Hospitals

Patients can be treated intramurally or as outpatients in general, university and special hospitals.

How the proposed reduction of beds to 3.7 beds per 1000 inhabitants will affect the elderly is not known. At this stage, it is not possible to judge whether this will lead to a reduction of the number of days spent in hospital or to a reduction of the number of patients. Certain assumptions were adhered to in the estimation of the policy variant. This led to the following results:

Table 3 Estimation of the number of days spent in hospital among the age group 65 years and older in the period 1982-2000 (total and per person) x 1000

	null variant		trend variant		policy variant
	abs.	d/p	abs.	d/p	abs.
1982	7,722	4.63	7,722	4.63	7,722
1985	7,995	4.62	7,827	4.52	-
1990	8,761	4.62	8,520	4.50	6,308
2000	9,664	4.63	(a)		6,429

a) see note 60 of chapter 2

d/p = number of days spent in hospital per person

There will be a limited increase of the number of geriatric wards in general hospitals. The object of this is to make diagnosis, treatment and referral of (very old) patients as effective as possible. Developments in geriatric outpatient departments will be important for the outpatient treatment of geriatric patients. It is assumed that in the future there will be an increase of these departments.

District nursing
Use made of district nursing per head of the population is stable (4.085 hours per elderly citizen per annum). Estimates in the null and the trend variant consequently coincide.

Table 4 Estimate of the number of hours of district nursing for the population of 65 years and older. Absolute and in number of hours per person

	Trend and null variant absolute	hrs/pat.	Policy variant
1979	6,529,000	4.085	In period 1984-86
1985	7,063,000	4.085	4% growth in vol.
1990	7,741,000	4.085	Norm: 1 dist. nurse
2000	8,534,000	4.085	per 2500 inhabs.

Policy aims at expansion of district nursing are in accordance with the objective of shifting the emphasis from intramural to extramural care. This will exert an influence on the nature of the aid provided by the district nurse. As a result of the restraints placed on the use of intramural facilities, there will be an extra increase in the demand for district nursing (substitution effects).
A recent development is the effort to make nursing aid available twenty-four hours a day.

Home help
More than half the aid provided by home helpers is to the elderly. In estimates of future use of home help, the null variant almost coincides with the trend variant, both absolutely and relatively. A reduction in growth is observable. What developments will take place in home help in the future is very unclear, since on the one hand a norm of 25.7 hours per annum for each elderly citizen living independently has been quoted, while on the other hand there can be only limited expansion in view of the limited financial resources.

Table 5 Estimate of annual use of home help in the period 1980-2000 (total number of hours of aid x 1000 and number of hours of aid per elderly citizen).

	null variant		trend variant		policy variant
	abs.	h/p	abs.	h/p	absolute
1980	28,553	17.70	28,553	17.70	-
1985	31,473	18.20	31,481	18.21	39,592
1990	34,221	18.06	34,137	18.01	43,393
2000	38,419	18.39	(a)		47,835

a) see note 60 of chapter 2
h/p = hours per person

The consequences of the limitation of admissions to homes for the elderly, nursing homes and hospitals, will also be felt in the field of home help.
In coming years there will be a need for considerable adjustment in the supply of aid (permanent availability, permanent supply of aid, and prompt availability of aid).

Housing for the elderly
Meeting the need for suitable housing for the elderly is an important instrument in the implementation of policy goals relating to the elderly. Suitable housing encompasses several forms with large differences in the level of facilities;

- ordinary small dwellings;
- adapted dwellings;
- monitored dwellings;
- service flats (in both purchase and rental sectors).

At the end of 1981 there was a shortage of nearly 110,000 suitable dwellings for the elderly in the Netherlands as a whole. Half of this shortage was to be found in the Western part of the country. Between 1980 and 1990 an increase of 10,000 suitable dwellings per annum is required, while between 1990 and 2000 an increase of 7500 per annum will be called for (estimates on the basis of a constant demand per age group with 1981 as base year).
For the period taken as a whole, an extra 2000 dwellings per annum will be necessary as a consequence of the restriction of admissions to homes for the elderly.
In view of the limited turn-over, the development of the future shortage will be mainly dependent on the number of newly constructed dwellings.

In addition to the more 'traditional' residential forms, in recent years there have been experiments with 'alternative' forms of housing (group dwellings, kangaroo dwellings). However, in view of the change in attitudes which these would necessitate, the reference scenario assumes only a limited expansion of alternative dwelling forms.

Co-ordinated work for the elderly
An important position is accorded to the co-ordination of social work affecting the elderly in the context of the so-called 'flanking policy'. This policy aims at deriving maximum benefit for elderly people from the relevant facilities. Emphasis is placed on lending support to initiatives of the elderly and persons with whom they are in contact, as well as to initiatives of volunteers. In view of the advocated policy for the elderly, it is expected that in future the importance of co-ordinated work for the elderly as part of the flanking policy will increase.

Summarizing, it may be said that in the reference scenario increasing de-institutionalization is envisaged. Nonetheless, if the supply of aid is not to fall short of demand, it will be essential to expand the amount of extramural aid provided. In addition to this, government policy aims at incorporating more informal aid (umbrella care). However, as was indicated above, there is evidence of social trends which cause considerable uncertainty to attach to perspectives for the increase of informal aid.

2.6 The economic context

In recent years there has been a sharp increase in the cost of health care and social services. Attempts at controlling costs have not resulted in the desired reduction of expenditure.
Whether in the future it will be possible to meet the costs of health care and social services depends not only on macro-economic developments (growth or decline), but also on the limits of care (how much of the national income may be expended on care, and how will priorities with respect to care develop?).

The reference scenario inculcates a limitation with respect to these questions. It has been established that in the period 1981-2000 there will be an increase of 0.6% per annum in collective expenditure if constant age-specific usage figures are assumed for the various facilities. In these calculations account is taken of both the decrease in the number of young people and the increase in the number of older people (medium variant population forecast). Assuming a constant ratio between national income and collective expenditure, it will be necessary for the national income to increase by 0.6% per annum to keep tread with demographic developments.

For an overview of the reference scenario see table 1.1 in chapter 1.

3 Medical and medical-technological developments

3.1 Introduction

The tremendous strides in medicine and in social services and facilities during the past 100 years have had the effect of doubling life expectancy at birth in industrialized countries. This first medical revolution has now come to a halt.
Given a combination of limited progress and sharp increase in costs, it is interesting to explore the future development of this relationship, and to review expectations with respect to medical science in general and to biomedical research in particular. Are developments already discernable which could be indications of a second medical revolution? And what would be the possible consequences for health and the volume and nature of health care?
To arrive at an answer to these questions, the scenario committee decided to expand its expertise by seeking the advice of thirty or so specialists from various branches of medicine. In May 1984 six meetings with experts in the area of medical and medical-technological research and medical practice were organized. These discussions were intended to give a more solid foundation to judgements in the scenarios regarding medicine and medical-technological developments between 1984 and the year 2000 than would have been possible on the basis of information from the literature and the knowlegde of the members of the scenario committee.
The general conclusion is that in the period till the year 2000 no essential medical breakthroughs are to be expected which would

result in any fundamental change in the present health situation and the system of health care. Neither are any breakthroughs to be expected in the treatment and curing of diseases in the short period till 2000. Developments which have already made their debut, especially in surgery, will benefit the elderly. Generally speaking, these developments will not have the effect of prolonging life as such, but will be especially valuable for diminishing disability, as a consequence of which the quality of life will be improved.
In view of the expectation that there will be no further breakthroughs in medical science in the immediate future, and the consensus of opinion on this matter among external experts, the committee has decided not to formulate alternatives for the variables 'medical and medical-technological developments' per scenario, but to treat them as one context.

3.2. Expectations with respect to medicine and pharmacology

It is expected that diseases with a long incubation period (including some types of cancer, coronary and arterial diseases, dementia, and diseases caused by persistant viruses) will continue to play an important role in the health of the population. The great disadvantage attaching to existing diseases with a long incubation period is that there will be little change in the impact of medical progress on the incidence of these diseases in coming generations. In view of the ageing of the population as a whole, this means that in the future there will be an absolute increase in the incidence of these diseases. For the present, no significant breakthroughs in cures or retardation of the progress of the most important diseases can be expected.
The emphasis for the future will be mainly on palliative measures, which for the patient will signify an improvement in the quality of life.

Expectations are limited with respect to the chance of cures resulting from breakthroughs in pharmacology. On the one hand, the pharmaceutical industry is dependent on insights from the field of medical science with respect to breakthroughs in the area of etiology. On the other hand, the pharmaceutical industry is also dependent on the economic remunerativeness of its discoveries, and will consequently not apply itself to seeking cures for every disease, but only for those for which investments in research and marketing outlets for new medicines promise to be profitable.

3.3 Expectations with respect to technology

The costs aspect

The costs factor plays a crucial role for the future development of technological apparatus. From a technical point of view, possibilities already exist for solving a great number of practical problems, but only at high cost. The question arises in how far these relatively limited improvements and refinements justify the costs involved in developing and applying this new apparatus. On the other hand, new technological apparatus may have the effect of reducing costs by cutting down on the amount of care necessary.

Medical-technical apparatus

It is expected that there will be improvements in scanning techniques for making diagnoses and for monitoring therapy. In addition to scanners, lasers will also gain in importance. Apart from this, however, no further improvements of any significance are expected in therapeutic apparatus.

Technical aids

As regards developments in the area of technical aids, it is expected that there will be considerable progress with respect to innovations particularly designed for the elderly (micro-electronics). These aids usually do not serve to prolong life, but considerably improve the quality of life of the (handicapped) patient, while also making it possible for him to continue to live independently in his environment. A large increase in the use of prostheses is expected. Expectations are especially high with respect to 'intelligent' prostheses. These self-regulating prostheses are simple to use, since they function independently of conscious effort on the part of the user by means of biofeedback systems. Another category of aids for which expectations are high are those suited for implantation. Examples of this are medicine reservoirs which, swallowed or implanted once, dispense uniform doses of the medicine and are effective for long periods.

Nursing

Nursing of the hospital patient in bed is exchanged for nursing of the patient out of bed. This could prevent a number of as yet common complications among the elderly.
Nursing will alter from passive nursing to nursing focused on activation of the patient, which will help to maintain the patient's condition at the highest possible level.

Information technology and epidemiology
As a consequence of the wide possibilities for the application of micro-computers, it is expected that there will be vast strides in information technology. In particular epidemiology is expected to profit from this trend as a result of the possibilities offered for collecting and analysing the enormous quantities of medical information. Thus insight could be obtained into the cause and spread of diseases, as well as the risk factors attaching to diseases. Objections attaching in the sphere of privacy should however be borne in mind.

3.4 Towards a second medical revolution?

From the very start, research strategy in medicine and pharmacology proceeded from the need to understand and control disease processes. Due to an often insufficient understanding of cells and cell processes, however, emphasis was often placed more on the control of sickness processes than on understanding them. During this period, chance played a not insignificant role in the solution of certain problems. The enormous improvement in the health of the population shows that this research strategy has been very successful. At present, however, progress leaves much to be desired. Nonetheless, this stagnation is not expected to be of a permanent nature. As a consequence of revolutionary developments in molecular biology, it has been possible to glean considerable insight into the structure and function of living cells. Thanks to this vastly increased knowledge of the healthy cell and the healthy organism, possilibities have been enlarged for describing, distinguishing and especially understanding pathological processes. The path has now been opened for medical science to adopt a more rational strategy. Expectations are that after the year 2000, the fundamental knowledge of cells, heredity etc., will have expanded considerably, and that fundamental changes in research strategy may then lead to a second medical revolution.

3.5 Expectations with respect to cell biology

In addition to research as to the causes and combating of specific diseases, fundamental research of the ageing process at the level of the cell plays an important role. Ageing involves a natural lessening and/or change of cell functions in the body, as a result of which the chance of contracting diseases increases. If it were possible to discover what exactly constitutes the process of ageing and what are the agents behind this process, it would become possible to distinguish the normal process of ageing from pathological ageing. In the context of the anticipated new research strategy, it will be possible to develop all sorts of drugs for curing or preventing pathological forms of ageing. This stage has however not yet been reached. Various problems are encountered in cell research with respect to the process of ageing. For instance no adequate animal models are available for research as to the ageing of higher human functions such as those of the brain.

Biotechnology

Hereditary information in healthy cells plays an important role in the ageing of the cell. This information is stored in coded form in the DNA, a component of the chromosomes of all cells. In the early seventies a fairly simple and rapid method was discovered for isolating the genes of nearly all living organisms and preserving them in an almost pure state and in almost unlimited quantities. It is expected that already before the year 2000 the human DNA will have been charted.

These developments offer enormous potential for application in medicine and pharmacy.

An already recognized application of the recombinant DNA technique is molecular cloning, involving the isolation of certain parts of the human chromosome and their implantation in a bacteria, which then produces unlimited quantities of the substance which has been genetically implanted.

For the future, this technique offers even more spectacular possibilities, including treatment of hereditary diseases. Nonetheless, numerous problems and unknown quantities still attach to the 'substitution' of genes for therapeutic purposes; which raises questions as to the realizability and desirability of application in the case of certain genetic diseases (even apart from possible risks).

Immunology

One of the body's most important defence mechanisms against foreign matter is the immune system. It is known that the process of ageing is accompanied by changes in the immune system, as a result of which the chance of contracting certain diseases is heightened. It is expected that it will become possible to steer the immune system. One application of this would be the combating of autoimmune diseases. The significance of this possibility for the elderly would however be small, in view of the limited number of elderly people

with pathological auto-antibodies. Immune deficiency is assumed to play no role in the development of cancer, but it must be taken into account in the combating of this disease, namely in the case of immunotherapy. Immunology is expected to contribute considerably to solving the problem of transplant rejection.
Moreover the application of knowledge of the immune system could be of great value in other areas such as for instance endocrinology.

Neurobiology
Research as to changes in neurons during the processes of ageing and dementia plays an important role in neurobiology. Greater emphasis will be placed on the relationship between 'normal' and 'pathological' ageing. This is especially important in connection with the possibility of developing a rational form of treatment for Alsheimer's dementia.

In view of the developments in cell biology, it may be concluded that in the near future enormous progress may be expected in fundamental biological research. Both as regards ethics and application this will have far-reaching consequences for medical science.

4 Scenario B: Increasing growth in demand for facilities

4.1 Introduction

This scenario proceeds from a larger growth in the demand for facilities by the elderly than would be expected on the basis of demographic developments (cf. the estimates with the future demand in the null variant of the reference scenario).

It is assumed that in general the demand for facilities will also be greater than might be expected on the basis of the trend variant of the reference scenario.

In the present scenario the increasing growth in the demand for facilities is brought about by a number of more or less consistent developments. These developments are seen as characteristic for a type of society in which achievement, labour, productivity and attachment to material assets constitute central values. The developments are more or less autonomous. Public health policy can exert little or no influence on them.

4.2 Social developments as they affect the health situation

The elderly of the future

The reference scenario described how in the future the elderly will be better educated and constitute a population group of increasing political significance. There will also be an increase in the (relative) prosperity of the group, especially as a consequence of the fact that more people receive a retirement pension and the higher level of pensions. It is assumed that these developments will have the effect of heightening the aspiration level and making them more assertive.

These points of departure also apply with respect to the present scenario. Further emancipation of the elderly is however hampered by the existing norms and values which apply with respect to the elderly in society (the struggle for emancipation).

The aspiration of the elderly to remain independent for as long as possible (an aspiration in which they have the support of government) nonetheless has drawbacks in addition to its many positive aspects. Deterioration of the physical condition and of certain functions (walking, sight, hearing) results in greater dependency on others. Acceptance of this dependency is a prerequisite for the positive experiencing of ageing.

In this scenario we proceed from the assumption that there will be a clear differentiation in the needs of the elderly. Some elderly people will desire to and be capable of participating to a high degree in social events, while others will, sometimes of necessity, participate only to a very limited degree.

Values and norms with respect to the elderly

It is assumed that there will be little change in the current discriminatory values and norms with respect to the elderly in society, which will mean that the wishes and desires of the elderly will be met only to a limited degree or not at all.

This will among other things find expression in the pressure placed on them to retire from the labour market at an early age, possibly to an even greater extent than is at present the case. Moreover the rate of unemployment among the elderly is fairly high. The growth of the labour supply till 2000 (especially as a consequence of participation in the labour market by married women and population growth in the past) will also exert a negative influence on this tendency. No part-time jobs will be created especially for the elderly. The official pensionage will remain stationary at 65. The discrepancies between the aspirations of the elderly and the manner in which these aspirations meet with acceptance in society will lead to a heightened degree of dissatisfaction among the elderly and consequently to heightened stress.

Volunteer work
The chances for the elderly to find paid work are thus viewed as limited in this scenario. This would create possibilities for performing more unpaid work (volunteer work). In this scenario an indication will be given of what forms of work lend themselves best for volunteers.

Decreasing solidarity
A decreasing degree of solidarity, both between the generations and within the generation of elderly, is one of the points of departure of the growth scenario.
Intergenerational solidarity is restricted by:
- the increasingly individualistic lifestyle;
- emancipation of women;
- geographical mobility.

Willingness to provide informal aid is closely associated with the type of aid required. The longer help is required, the less the chance of obtaining informal aid. Moreover the elderly often show a preference for professional aid.
The effect of the individualistic lifestyle on intergenerational solidarity (among the elderly themselves) is that the elderly will demand the right to professional aid both for themselves and for their peers. An exception will however be made with respect to persons belonging to their own household (in most instances a partner).
The traditional nuclear family will continue to be the predominant form of household encountered. Alternative households (communes, kangaroo dwellings) will not take hold to any significant degree.

Attitudes with respect to sickness and health
Decreasing responsibility for one's own health is one of the characteristics of the growth scenario:
- the earlier-mentioned stress reactions (i.a. as a consequence of the high level of unemployment or premature pensioning) will

lead to an increase in risk behaviour (smoking, drinking);
- responsibility for health (inclusive of the health problems brought about by stress an dissatisfaction) will be increasingly laid at the door of the physician.

Till the year 2000, this tendency will however hardly have any implications for the health of the elderly.
In this scenario there will be an even greater increase in tolerance with respect to euthanasia than in the reference scenario.

Increasing professionalization
The increasing tendency towards professionalization in health care and social aid continues. Pressure for increased professionalization comes from two sources.

(a) Pressure from professionals
This may be ascribed to two developments:
- Improvement in possibilities for making diagnosis and the administering and monitoring of therapy thanks to new technological apparatus, as well as the development of all sorts of technical aids.
- The development of new specialisms; three examples are discussed.

(b) Pressure from the consumer
The emancipated, more assertive, and well-informed older people of the future, will want to make use of the best facilities available at a certain moment. They will take steps to avoid or to dispense with restrictions with respect to treatment.

4.3 Consequences for facilities

The greater possibilities on a medical and medical-technological level, as well as the high inclination to consumption of health facilities and the lessening of informal aid, will in general lead to a sharp increase in the demand for facilities, both absolutely and relatively. This increased demand will however not apply equally to all facilities.
In the growth scenario a qualitative estimate is made of the consequences for (health) care facilities. Exceptions to the pattern of growth are homes for the elderly (there will be both an absolute and a relative decline in demand for this facility) and the nursing home (increase in demand for this facility runs parallel with demographic developments).

4.4 The economic context

By definition, economic growth is essential for this scenario. A higher growth in national income will be required than in the case of the reference scenario. The increase in demand will be greater than might be expected on the basis of the (double) ageing alone.

See table 1.1 in chapter 1 for an overview of the growth scenario.

5 Scenario C: Decreasing growth in demand for facilities

5.1 Introduction

In the 'shrinkage scenario', a picture is given of a number of more or less consistent developments which, provided there is a corresponding development in supply, could result in a decrease in the (growth of) demand for facilities.

Various trends (emancipation of the elderly, an increasing level of education, more tolerant attitudes with respect to passive or active euthanasia, increasing political significance of the elderly) are identical with those encompassed in the reference scenario and the growth scenario.
In accordance with the trends indicated in chapter 3, neither are any significant shifts expected in this scenario as regards the health situation and medical and medical-technological developments. Though there will be shifts in attitudes with respect to health and disease, lifestyles and socio-economic position, the influence exerted by these shifts on the health situation of the elderly till the year 2000 is expected to be marginal. Over a longer period of time the consequences will probably become more visible.

The most important trend which in this scenario determines the decreasing pressure on facilities is thus not the (improved) state of health of the elderly, but changes in the conception of care.

5.2 Social developments

Altered attitudes with respect to illness and health

The importance attached to good health will increase. This attitude fits into a society type in which immaterial values (health, relationships, self-improvement) are considered increasingly important.

Though perhaps only over a longer period of time, the incidence of disease and premature death will be reduced as a consequence of an increasing feeling of responsibility for one's own health and a lessening of risk behaviour. There are however limits to the extent of which the individual can be expected to bear responsibility for his own health, since this responsibility relates only to 'self-imposed risks'. Other factors such as conditions of work, environmental pollution, food production etc. also influence health. High priority is attached to primary prevention. A longer period of activity among the elderly may contribute also to diminish morbidity and mortality.

Work and retirement

There will be a reversal in the trend towards an increasing period of inactivity among the elderly brought about by premature retirement and increased life expectancy. Partly thanks to gradual or flexible pensioning (between the ages of 60 and 70) and rigorous shortening of working hours (to 25 hours a week); there will be an increase of participation by the elderly in the labour process. The retirement question is being studied from various aspects (economic-demographic, ethical). Various variants of income composition of the elderly are examined. There will also be an increase in participation in volunteer work for and by the elderly. There will be a further decrease in the difference in evaluation of paid and unpaid work.

As a consequence of this, the view that the elderly constitute an economically inactive category will disappear.

Long-term consequences for health

According to this scenario, over a longer period of time the health of the elderly could improve as a consequence of:

- a reduction of risk behaviour in combination with collective measures for the furtherance of health;
- a reduced individual work load and better distribution of jobs between the sexes;
- improved adaptation to individual wishes (i.a. through flexible pensioning), possibly resulting in a heightening of well-being;

- an improved income position (under certain conditions) and
- decreased unemployment.

As a consequence of these developments it is expected that there will be a lessening in the difference in life expectancy between men and women. This decrease will be larger than that already found (namely a half year) in the reference scenario. This will have interesting long-term consequences for the umbrella care with which partners provide each other.

5.3 Developments relating to facilities for the elderly

The system of facilities for the elderly will be adapted to the changing concepts of care. A greater diversity of residential and care facilities will become available. In this context the importance of informal aid will increase. A combination of these factors will have the effect of reducing the demand for especially intramural facilities, if not absolutely at least relatively.

Central points of departure
The altered views with respect to the concept of care are characterized by the following phenomena:

- reduction of the increasing professionalization in certain areas of care;
- a limitation of the admission of the elderly to the intramural sector (de-institutionalization);
- diversification of care (expansion of the choice possibilities and 'cut to measure' care);
- reduction of the tendency to label all problems as medical (scientific purity in medical practice);
- upgrading of the concept 'care' (especially of importance as a consequence of the increase in chronic-degenerative diseases).

These developments have consequences for all facilities. The greatest changes will be found in facilities with a residential function or a nursing function.

The home for the elderly in its present form will disappear. In the course of time one intramural small-scale long-stay facility will come into being for old people who require a high degree of assistance (a combination of home for the elderly and nursing home). These long-stay facilities will at the same time function as headquarters for aid services to elderly people living independently (district function). The range of care and living arrangements for

elderly persons who either require only little assistance or can manage on their own will increase (the traditional family, open families, LAT relationships, concubinage, communes, monitored dwellings).
It will continue to be possible to obtain additional outside aid (either professional or informal). The emphasis will be on forms of solidarity within the generation of elderly.
Aid from children will not increase.
The 'shrinkage scenario' furthermore presents the consequences for those facilities where nursing and medical treatment fulfill a central role (nursing homes and hospitals). It is assumed that there will be no further increase in the number of geriatric wards in general hospitals, though the specialism geriatrics will not disappear. Technological skills with respect to diagnosis and therapy will be applied selectively.

Finally, possibilities for self-treatment will increase as a consequence of the availability to the public of diagnostic apparatus and computerisation ('telediagnosis').

It is concluded that a (relative) decrease in demand will be observable only with respect to intramural facilities. Both a relative and an absolute increase in demand for extramural facilities is expected.

5.4 The economic context

When this scenario is confronted with assumptions concerning economic growth or decline, two variants are obtained. In the event of economic growth in combination with a high level of unemployment, a basic income will be introduced. In the event of economic decline, solutions will be sought in the direction of the introduction of a (limited) public service obligation and raising the level of unconditional income transfers to the elderly. Though this increase will be met from public funds, this will probably be amply compensated by the resulting reduction in the use of facilities.

See table 1.1 in chapter 1 for an overview of this scenario.

6 Disturbing developments

6.1 Introduction

Two disturbing developments are discussed in the scenario report:
(a) postponement of dementia by five years;
(b) an extreme lessening of intergenerational solidarity (care of parents by children).
The first 'critical incident' relates to medical and medical-technological developments and could result in a reduction of demand for (health) care facilities. The second incident relates to social developments and will result in an increased demand for professional assistance. Both incidents are analysed not only qualitatively, but also quantitatively.

6.2 Postponement of dementia

Five reasons are put forward for the choice of postponement of dementia:
- dementia is a syndrome encountered predominantly among the elderly;
- dementia results in a high degree of impairment and tarnishing of the personality;
- incidence and prevalence of dementia increase with age. Double ageing consequently means both an absolute and a relative increase of dementing old people in the age category 65 years and older;
- the increasing number of dementing old people increases the pressure on intramural facilities;
- life expectancy of dementing patients is lower than for the elderly taken as a whole.

Epidemiological aspects
Dementia is not a clearly defined disease syndrome. Consequently the prevelance and incidence figures for dementia differ widely for different countries. Moreover methods of data collection are also responsible for these differences.

With respect to the Dutch situation epidemiological data are scarce except for old people in the intramural sector. For the purpose of calculations of postponement of dementia, use was made of figures deriving from nursing homes.

Data from other countries show that the prevalence and incidence figures must be differentiated according to sex and age. A combination of data from various studies shows a prevalence among men of 3.9% (65-69 years of age) to 13.2% (80 and older) and among women of 0.5% (65-69 years of age) to 20.9% (80 and older). The data relate to the 'chronic brain syndrome' and concern only the categories 'seriously' and 'moderately' demented patients. Application of these data to the Dutch population of 65 years of age and older results in an estimate of 117,000 dementing old people (44,000 men and 73,000 women, namely 6.4% and 7.8% respectively).

Medical and medical-technological developments

It would appear that till the year 2000 no significant breakthroughs are to be expected with respect to knowledge of the causes and processes leading to dementia. Research as to therapeutic possibilities will in the future continue along present lines:

- therapy at a pharmacological level (neuropeptides, neurotransmitter normalization);
- social or environmental therapies (structured stimulation, reality orientation therapy, enriched environment);
- alterations in the dietary pattern (eradication of vitamin shortages);
- transplantation of embryonic brain tissue.

There will be an improvement in possibilities for early and differential diagnosis of dementia as a consequence of:

- more attention being devoted to dementia in the training of physicians;
- better education of the public on this matter;
- improved technological possibilities (scanners such as PET scan and NMR);
- an increased degree of multidisciplinary co-operation in the observation of patients (tracking down of reversible (pseudo) dementia);
- further expansion and intensification of ambulatory mental health care.

Towards a calculation model

Calculation of the consequences of the above-mentioned developments for the prevention of dementia and the demand for facilities for these patients till the year 2000 is no easy matter.

The best possibilities probably rest in improved diagnosis and perhaps also certain therapeutic measures.
With respect to the nursing home sector a rough estimate has been made of the consequences of a five-year postponement of dementia. It is assumed that the age-specific admission figures for dementia will move upward by one five-year cohort. If such a shift were realized by the year 2000, this would mean a reduction of one-quarter in the number of nursing home patients (compared with the situation in the year 2000 without such postponement). As compared with the situation at the beginning of the eighties, postponement of dementia would mean an increase of 9% in the number of dementia patients in nursing homes in the year 2000, while if there is no postponement of dementia, the number of dementing patients in nursing homes will increase by 48% by comparison with the situation at the beginning of the eighties. It should be stressed that these results are of only limited value in view of the various assumptions employed in the calculation model.

6.3 Extreme decrease of intergenerational solidarity

Scenarios B and C already assume some decrease of intergenerational solidarity. What will be the consequences of the extreme situation which will come about if in 1990 or in 2000 the aid provided by children to their parents suddenly disappears? In reality this development will not be so extreme, but it is by no means unrealistic to proceed from the assumption of a decreasing level of umbrella care in view of the following factors:

- a decline in the number of children and the fact that they leave home at an earlier age;
- increasing emancipation of women (decreasing willingness and ability to provide umbrella care);
- an increasing tendency to individualization;
- increasing geographical mobility;
- the growing objection of the elderly to the asymmetric relationship with providers of informal aid and a consequent preference for professional aid.

Basic data with respect to the aid relationship children - parents
Data for making these estimates were derived from the "Living Conditions Survey" carried out in 1976 by the Central Bureau of Statistics among the Dutch population of 55 years of age and older. In this survey a distinction was made between elderly people living independently and elderly people in homes. This latter category is not included in the calculation model, since in their case it is

very unlikely that children are involved in providing care. With respect to the former category, account was taken only of the relationship between parents and children living away from home. The reason for this is that no data are included in the CBS survey as to why children continue to live with their parents.

10.3% of elderly people in the age category 55 years of age and older who live independently, regularly receive assistance from children living away from home. 15.4% of people in the age category 65 and older receive such assistance. Data are also available regarding age category and sex. Together with domestic personnel privately engaged by the elderly, children constitute the most important group of providers of aid (compared with other family, neighbours, acquaintances, volunteers, professional home helps and district nurses).
67% of children provide aid for 1 - 6 hours a week, while 18.9% provide more than 6 hours of aid a week.

Towards a calculation model
The calculation proceeds from the assumption that only the umbrella care provided by children living away from home will alter, and that all other conditions will remain the same.
It has been estimated that in 1990 there will be 124,000 men and 216,000 women who receive aid from children not living at home, while in 2000 these numbers will have risen to 142,000 men and 238,000 women (this estimate is based on the number of independently residing people of 55 years of age and older). The disappearance of this umbrella care would imply that these elderly peoply will have to seek this essential aid elsewhere. We can only speculate on the directions in which and degree to which the demand for aid will develop. One possibility has been calculated: other assumptions will yield other results.

We proceed from the following assumptions:
- less than 6 hours aid per week. 50% of the people involved will appeal to first echelon facilities, 40% will fall back on paid domestic aid and 10% will move to homes for the elderly.
- more than 6 hours aid per week. 50% move to homes for the elderly, 40% appeal to first echelon facilities, and 10% fall back on paid assistance.
- number of hours of aid unknown: 50% appeal to first echelon facilities and 50% call in commercial assistance.

According to these assumptions, in 1990 or 2000 the following increases may be expected in the demand made on the various facilities:

Table 6 Estimated increase in the demand for facilities in the period 1990 to 2000 (to the nearest thousand) in number of persons

	1990	2000
homes for the elderly	32,000	36,000
dwellings for the elderly	23,000	25,000
first echelon facilities	164,000	183,000
commercial aid	122,000	136,000

6.4 Disturbing developments in relation to the scenarios

The effects of the postponement of dementia on the use of facilities should not be overestimated. The possible reduction in the number of nursing home patients will probably not result in a reduction of the demand for beds. At the moment, there is a shortage of psycho-geriatric beds according to the present norm, while moreover raising of the norm is under discussion. Furthermore, the effect will in practice be less on account of the often encountered multimorbidity (on the basis of the secondary diagnosis the patient still needs to be admitted to a home) plus the chance that other diseases make their appearance in the years gained.

By comparison with the trend variant, the reduction in the amount of umbrella care provided by children for instance implies an increase of 22% in 1990 in the use made of homes for the elderly. By comparison with the null variant this means an increase of 19%.

The disappearance of intergenerational solidarity has less dramatic consequences for scenarios B and C, since both scenarios already to some extent take account of a certain lessening of this solidarity. In scenario B this would mean extra pressure on professional facilities, while in scenario C the reduction in the (growth of) demand would be counteracted by this disturbing development.

7 Application possibilities for scenarios

7.1 Introduction

Scenario reports yield maximum returns if used as tools, for instance by adding to them or carrying out supplementary calculations.

7.2 Scenarios as 'learning environment'

Scenarios may serve as 'test situations' for policy proposals. By confronting policy proposals with favourable and unfavourable 'contexts' it is easier to foresee what will be the effects of new policy. After each successive confrontation, the policy proposals can be adapted till they have proved adequate to lay before the decision-making authority. Some developments in health care in Finland are mentioned in this report as examples of such a learning process. One of the problems which arise in this context is getting processes of change started and maintaining this change (for instance as regards prevention and lifestyle). Successful influencing campaings as a rule show certain characteristics. The reader is referred to the literature relating to 'planned change'.

7.3 Scenarios and (health) care facilities for the elderly

Three patterns for health care facilities for the elderly were dealt with in scenarios A, B and C, namely: 'maintaining present course' in A, 'top care' in B and 'towards a reciprocal aid society' in C.

The scenario report goes deeper into the consequences of other combinations between patterns of facilities and context scenarios (see figure VII.1 in chapter 7):

- 'maintaining present course' and scenario B (increasing growth in the demand for facilities) will result in serious problems: there will be waiting lists; considerable pressure will be placed on the ability of the elderly to manage for themselves; for some of the elderly commercial facilities may be a solution;
- in the combination 'maintaining present course' and scenario C

the pattern of facilities would have to be adapted to the decreased, but altered, demand (small-scale facilities). It is not unlikely that initial problems will be encountered;

- in the pattern of top care under the conditions which have been sketched in the reference scenario, some of the elderly will be excluded from the health care facilities; there will be waiting lists, while rates and premiums will increase.
  Part of the unsatisfied demand may take recourse to commercial facilities;
- top care in the event of a declining demand for health care facilities is theoretically possible; few people would make use of the top care; the emphasis is on reciprocal aid, a change in mentality and alteration in lifestyles;
- a combination of conditions from the reference scenario and the pattern 'towards a reciprocal aid society' would - apart from initial problems - exert a positive influence on the health care of the elderly;
- finally there is the combination of increasing demand for facilities and a pattern of facilities which paves the way for a reciprocal aid society. These two developments are difficult to reconcile; it is more likely that they exert a negative influence on each other.

It may thus be concluded that the context scenarios in combination with the various patterns of care yield a number of negative side effects.

7.4 A closer look at putting the scenario report to active use

The scenario report could lend support (a) in the assessment of policy memoranda (i.a. the Policy Memorandum on Health in 2000), (b) in discussions with respect to the range and type of investments in medical and in social-scientific research relating to ageing, and (c) in case of the introduction of an 'early warning' system.

7.5 Concluding remarks

The scenario project on ageing has made clear that Dutch policy with respect to health (care) of the elderly will have to be prepared for autonomous developments of a very diverging nature.
It proved possible to provide indicators for the choice of a pattern of health care facilities for the elderly to be strived after, while the consequences of alternatives were clarified.

1 Introduction

1.1 Background

In the Explanatory Memorandum accompanying the Government Budget 1983 the Minister of Welfare, Health and Culture stated that it was his intention to intensify and restructure research as to long-term developments in the sector of health care. In the framework of the implementation of this policy, the Secretary of State for Welfare, Health and Culture, J.P. van der Reijden, on the 9th of March 1983 installed the Steering Committee on Future Health Scenarios.
The task of the Steering Committee is to advise the Minister and Secretary of State on the future of health and health care in the long-term in the Netherlands, with a view to enlarging the anticipatory capacity of policy.

Preparations by the Minister of Health for the intensification and restructuring of research as to long-term developments in health care included among other things consultation with several experts in the sector of health care. Expanding on the results of these preparatory activities, around the middle of 1983 the Steering Committee set up four Scenario Committees. These committees were entrusted with the task of drawing up scenarios with respect to (a) coronary and arterial diseases, (b) malignant growth, (c) life-styles (especially those involving risk) and (d) ageing. Each of these four committees is aided in its task by a research team.

The objective of these scenarios is to provide the long-term data required as a basis for the Policy Memorandum on Health in 2000 which the Secretary of State will place before the Second Chamber in the near future.

Furthermore, from the start, one of the goals was that the reports of the scenario committees should serve as part of the Dutch contribution to the work of the World Health Organization. In 1979 WHO launched the programme 'Health for All by the Year 2000'. Future research, management and planning of health care play an important role in this programme. Three countries, namely Finland, Sweden and the Netherlands, have undertaken to carry out trial projects with the scenario method. On the basis of the results of these trial projects, further steps will be taken to enlarge on the scenario method in the interest of this WHO programme.

1.2 Design and course of the scenario project on ageing

This report contains the scenarios developed jointly by the scenario committee on ageing and the research team. The scenario committee on ageing consists of:

- Prof. Dr. C.F. Hollander, Chairman; Director of the Research Institute for Experimental Gerontology, TNO, Professor Extraordinary of Medical Gerontology, State University of Utrecht;
- Prof. Dr. W.J.A. van den Heuvel, Professor of Medical Sociology, State University of Groningen;
- Dr. P.C.J. de Koning, Economist, Organization Advice Bureau Verlinden-Wezeman, Oosterbeek;
- Dr. A. Fuldauer, Medical Director, Eugeria Nursing Home, Almelo;
- Dr. D. van der Meer, Director of the Hospitals Institute of the Netherlands, Utrecht;
- Prof. Dr. P.J. van der Maas, Professor of Social Health Care, Erasmus University, Rotterdam;
- Prof. J.J.M. Michels, GP in the Nursing Home Kalorama, Beek near Nijmegen, Professor Extraordinary of Nursing Home Medicine, University of Nijmegen;
- Dr. D.J.B. Ringoir, Medical Chief Inspector for Mental Health, Leidschendam.

The research team, which derives from the Research Group for Planning and Policymaking of the Department of Social Sciences of the State University of Utrecht, consists of:

- Prof. Dr. H.A. Becker, Professor of Sociology, Co-ordinator;
- A. Klaassen-van den Berg Jeths, Sociologist;
- A. Kraan-Jetten, Sociologist;
- R.J.T. van Rijsselt, Sociologist.

The scenario committe on ageing has formulated the following problems in close co-operation with its research team:

(1) What are likely to be the most important (future) developments which will exert an influence on the health situation of the elderly in the Netherlands in the period 1984-2000?

(2) In view of the future health situation of the elderly and their increasing share in the Dutch population, what are the possible patterns of (health) care facilities in the period 1984-2000?

The first question relates to developments which take place more or less autonomously of the various actors in Dutch health care; developments thus on which these actors can exert little or no influence. By 'actors' are meant the Ministry of Welfare, Health and Culture, the Health Insurance Fund Council, professional medical associations, medical faculties etc. These autonomous developments take place partly within the Netherlands and partly in other countries. As far as other countries are concerned these developments relate mainly to innovations in medical science and in the care and treatment of patients. As far as the Netherlands are concerned developments relate mainly to national economic developments, developments in household incomes, or future energy supply.

The second question relates to the patterns of (health) care facilities which are theoretically within reach in this country.

The second question also defines the limits of the present scenario project. The choice of the pattern of health care facilities, as well as the choice of strategies for realizing this pattern will not be made in the present project, but will be set out in the Policy Memorandum on Health in 2000.

The age of 55 has been chosen as the lower limit for the category of elderly persons. The choice of this **age limit** is motivated by the fact that around this age people experience important changes in the pattern of their lives, such as premature retirement from the labour process, the fact that the youngest child leaves home, etc. The second argument for adhering to the age limit of 55 years of age is that many of the diseases from which the elderly suffer begin to make themselves manifest around this age. Adhering to this age limit could have the effect of focusing greater attention on early diagnosis of these illnesses.
A further argument for the choice of this age limit is the fact that many (statistical) publications employ this age limit.[1)]

On the other hand, it may be argued that the demand for aid as a consequence of illness and invalidity becomes clearly manifest only

at a much later age, especially around the age of 70. For this reason the age limit of 65 years of age will also be employed where the use of (health) care facilities is discussed.
The fact that WHO in general employs the age limit of 65 years of age is a further contraindication for the use of the age limit 55 years of age.

Persons accruing to the category 55 years of age and older are by and large quite capable of leading an independent and healthy life. A lot of these people are still completely or partially active in the labour process. Problems of illness and invalidity encountered among the elderly relate only to some of these people; moreover, many complaints are often concentrated in one person. Problems of social isolation or limited independence also only relate to some of the people in this age category. The scenario project on ageing has consequently been carried out with the conviction that there is nothing to be gained by generalizations on 'the problems of the elderly'.

The year 2000 has been adhered to as the **time horizon** for the scenarios. In view of the necessity for attunement with the Policy Memorandum on Health in 2000, this would seem an obvious choice. Certain developments are however viewed in a wider perspective in the scenarios. It is for instance possible to obtain a relatively reliable picture of demographic developments over a longer period of time. For some developments extrapolations till the year 2030 are even possible.

Developments in and concerning health care which play a role between 1984 and 2000 lend themselves only to a limited degree for forecasts which are reliable and precise enough for use as a 'hard' basis for policy decisions. Where possible, especially in the sphere of demographic developments, forecasts are used. Other developments which lend themselves less for the making of forecasts are also included in the scenarios.

The first step in the present scenario project was the carrying out of a background study. This study analyses the most important aspects of the history of health (care) problems of 1984 and preceding years. With respect to some subjects the study goes back to before 1970. The report of the Background Study, which was published in its definite form in 1984, consists of the following parts:

(a) A scheme shows the most important clusters of variables which influence the question of health (care) for the elderly in the Netherlands. This scheme adheres closely to the system to be employed in the Policy Memorandum on Health in 2000. This scheme is shown in Figure 1.1.

(b) With respect to each cluster of variables an estimate has been made of what developments have taken place between 1970 and 1984, and what are the bottlenecks involved.

(c) Where possible and desirable statistical data have been analysed for each cluster of variables.

The Background Study can be employed to place the scenario which have been designed in this project in a wider framework. The report of the Background Study has been written in such a way that on the basis of this report readers design their own scenarios. Designing new scenarios will sometimes involve the collection of additional data.

Around the middle of 1984 the scenario committee on ageing and the research team organized six consultations with groups of experts from the field of medical and medical-technological research and medical practice. These group discussions were intended to give a wider foundation to the pronouncements in the scenarios on medicine and medical-technological developments between 1984 and 2000 than was possible from the literature and the knowledge of the members of the scenario committee. Appendix I of this report includes a list of the names of participants in the consultations. Appendix I also gives a summary of the results of the consultations.

After several rounds of designing, adjusting and redesigning, three scenarios were ultimately drawn up. These scenarios show possible 'contexts' in which (health) care facilities for the elderly in the Netherlands can be realized between 1984 and 2000.

During the implementation of the present scenario project there was regular discussion with the Steering Committee, the secretariat of the Steering Committee and with the other scenario teams. The object of discussions was inter alia to achieve the necessary attunement of the four scenario projects. It was found that due to their concentration on specific types of diseases, the scenarios on coronary and arterial diseases as well as those on cancer called for a specific approach. The scenarios on ageing bear most relationship to those on lifestyles.

Figure 1.1 Scheme of clusters of variables from the Background Study

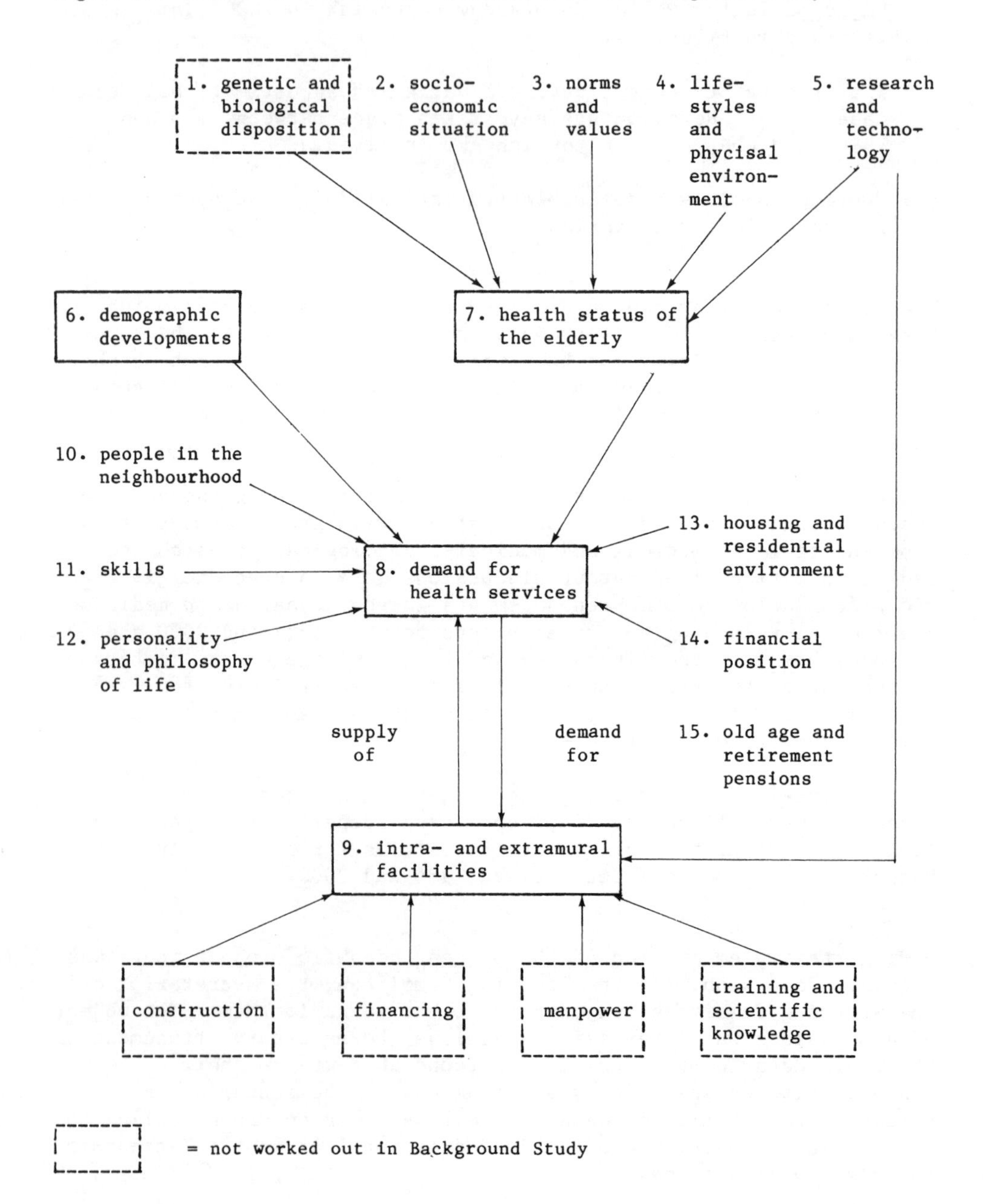

A preliminary overview of the results of the scenario project on ageing was presented and discussed at a conference of WHO held in Scheveningen in August 1984.[2]

1.3 Taking a further look at scenarios

In the first place, it would seem desirable to give a definition of the concept scenario. A wide definition as: "A scenario is a description of the present situation of a society (or part of it), of possible or desirable future situations of the society, as well as of series of events deriving from the present situation which could bring about those future situations".[3] This definition calls for clarification.

The clause "A description of the present situation of a society (or part of it) ..." refers to the Background Study. Such an investigation of the trends and bottlenecks is an indispensable part of each scenario. Moreover the depth of the background study is determined by the specific nature of a certain scenario project. Where mention is made of 'possible future situations', expected 'contexts' for policy are meant. A policy-maker nowadays often has various possible 'contexts' drawn up, so that he can test the applicability of each of the strategies he considers in each of the possible future contexts.[4] The policy-maker is thus enabled to choose his ultimate strategy in such a way that he is prepared for the worst. In this report these 'contexts' will be indicated further as 'scenarios' (in a narrower sense).

'Desirable future situations' may be envisaged as future situations which are considered to be both worth striving after and realizable. They constitute the potential goals from which the policy-maker can choose.

The 'series of events' consists partly of processes which could lead to each of the contexts, and partly of the policy activities which could lead to realization of the 'desired future situations'.

In the scenarios presented in this report, the focus is on the 'possible future situations'. The report offers especially the most important possible 'contexts' within which till the year 2000 (health) care facilities in the Netherlands must be realized. The 'contexts' consist mainly of developments, which, viewed from a vantage point of the actors in Dutch health care, take place 'autonomously'.

The contexts as it were provide the actors involved with a 'learning environment'. A learning process can derive from 'testing' the various policy resolves with respect to various 'possible future situations': gradually improving policy resolves and attuning these resolves to circumstances which fall short of or exceed expectations. Health care policy in the Netherlands would also do well to 'hope for the best but prepare for the worst'. Proceeding from these considerations, three (context) scenarios are presented in this report.

The first scenario presented is the **reference scenario**. This picture of the future proceeds from a health situation of the population which is identical with the present one, so that no significant trend deviations have been incorporated. In other words, the health situation and the use of health care facilities are influenced mainly by demographic developments and by developments observed in the use of the facilities. The reference scenario serves as a screen against which the other scenarios can be viewed.

On the basis of information from experts, **medical and medical-technological developments** till the year 2000 are indicated fairly unequivocally. This has led to the decision not to differentiate this component in the growth and shrinkage scenarios.

Next follows the **growth scenario**. As opposed to the reference scenario, this scenario proceeds from a significant increase in the demand for health care facilities, an increase which is larger than might have been expected on the basis of the calculations in the reference scenario. In this scenario developments are discussed which could lead to this state of affairs, such as far-reaching professionalization of health care, a greater inclination to consumption of facilities, reduced inter and intragenerational solidarity.

The last scenario is the **shrinkage scenario**. This is presented as the antithesis of the growth scenario. The shrinkage scenario proceeds from a decreasing (growth of the) demand for facilities.

Developments which could lead to this are discussed, such as a decreased inclination to consumption of facilities brought about by a heightened feeling of responsibility for one's own health, a greater degree of solidarity among the elderly, etc.

Table 1.1 gives an overview of the three scenarios.

Are these three (context) scenarios all equally likely? On the basis of developments in past decades, the growth scenario would appear to be the most likely. Second in line is the reference scenario. The shrinkage scenario brings up the rear. This order of likelihood however does not lessen the value of the three scenarios as 'learning environment'. The strategy for (health) care policy till 2000 which is ultimately chosen must be realizable in each of the three possible future situations. The strategy must be capable of standing up to extreme tests. A certain 'stylization' or 'overaccentuation', intrinsic to the use of the scenario method, is precisely desirable for situations in which policy cannot be prepared mainly on the basis of forecasts.

Table 1.1 Overview of the scenarios

ASPECTS	SCENARIO A	SCENARIO B	SCENARIO C
demographic development	medium variant CBS forecast	medium variant CBS forecast	medium variant CBS forecast
state of health	unaltered in relative sense, in abs. sense deterioration as consequence of demographic effect	deterioration in relative sense, but only after 2000	improvement in relative sense, but only after 2000
gap in life expectancy between men and women	slight decrease in accord. CBS forecast	slight decrease in accord. CBS forecast	greater decrease long-term (after 2000)
medical-technological development	unchanged	increase in possibilities; no spectacular breakthroughs	increase in possibilities; no spectacular breakthroughs

social developments	SCENARIO A	SCENARIO B	SCENARIO C
. educational level of elderly	increase	increase	increase
. emancipation elderly	increase	increase	increase
. number of children (re. umbrella care)	decrease	decrease	decrease
. emancipation of women (re. umbrella care)	increase	increase	strong increase
. tolerance of euthanasia	increase	strong increase	strong increase
. income	improved pensions	improved pensions	more pension recipients; slightly higher pensions
. behaviour and attitudes to health	unchanged	increase in risk behaviour i.a. due increased stress	greater responsibility for own health
. labour	constant relative number of older people in labour process	relatively less older people in labour process	shortened working hrs.; more older people in labour process
. retirement	unchanged	increase in no. of prematurely retired persons	flexible/gradual
. intergenerational solidarity	decrease	decrease	decrease

social developments (continued)	SCENARIO A	SCENARIO B	SCENARIO C
. intragenerational solidarity	unchanged	decrease	increase
. alternative living arrangements	to a limited degree	decrease	increase
. direction in which attitudes regarding health care develop	de-institutionalization; more umbrella care desired	professionalization; more medical intervention; emphasis on early diagnosis and cure	strong de-institutionalization; de-professionalization; altered opinion of sickness; diversification; upgrading of care

demand for facilities	SCENARIO A						SCENARIO B		SCENARIO C	
. general	null variant		trend variant		policy variant		increase greater than trend and 0-variants		decrease in (growth of) demand for esp. intramural facilities	
. per facility	abs/rel[5]		abs/rel		abs/rel		abs	rel	abs	rel
GP care	+[6]	0	+	0	+	+	+	+	+	0
amb.ment. health care	+	+			+	+	+	+	+	0
homes for elderly	+	+	+	-	0		-	-	)	
									) -	-
nursing homes	+	+	+	+	+	acc. to norms	+	0	)	
day treatment	+	+	+	+	+		+	+	+	+
hospitals	+	0	+	-	-		+	+	-	+
district nursing	+	0	+	0	+		+	+	+	+
home help	+	+	+	+	+		+	+	+	+
home for elderly	see section 2.5.9								+	+
co-ord. work for elderly	see section 2.5.10									
economic context	null variant: annual growth of Nat. Inc. of 0.6% essential						shrink/growth		shrink/growth	

The three scenarios present future situations and processes which, in spite of overaccentuation, are characterized by 'high probability - high impact'. Future situations will however also be influenced by developments which may be characterized as 'low probability - high impact'. It would be possible to mention a large number of possible developments in this category. Two such **disturbing developments** are further analysed in this report. The two developments chosen are a five-year postponement in the appearance of dementia, and a further decrease in intergenerational solidarity, with the result that children are no longer prepared to care for elderly parents. In both instances the research team have estimated what would be the resulting increase or decrease in the demand for facilities if these two developments should come about.

As far as **possible health care facilities** for the elderly are concerned, **three patterns** have been designed. The first pattern consists of a projection of policy with respect to facilities as carried out in 1984 and preceeding years. This would in fact simply mean **maintaining the present course**. Policy changes which were decided on in 1984 will be put into effect in this pattern.

The second pattern entails **top care**, however only in a strictly medical and medical-technological sense. All patients receive the best of the best, the newest of the new as seen from a point of view of the 'medical model'. Such a model of medical and medical-technological top care is conceivable in the case of increasing prosperity. The system is however also conceivable in case of a constant or decreasing degree of prosperity, for instance by means of raising fees or tolerating long waiting lists. In this case such a pattern would only be available to a select group.

The third pattern aims at a **reciprocal aid society**. In this pattern, the elderly are stimulated to manage for themselves, to keep themselves occupied, as well as to assist each other and participate actively in society. Health care facilities shift their priorities in such a way that 'care' receives somewhat more attention than is at present the case. Where possible small-scale facilities are stimulated.

In the patterns of health care facilities too a certain degree of 'stylization' and 'overaccentuation' is applied. For the purpose of testing these patterns in different contexts, such overaccentuation offers more advantages than disadvantages.

How can contexts and patterns of facilities be employed in the preparation of policy? Figure 1.3 presents a system which could be used for this purpose. Each of the patterns is considered in turn.

It is then assessed in how far the pattern in question could be realized in each of the three contexts, possibly with disturbing developments. This is discussed in chapter 7.

Figure 1.3 Scheme for viewing the relationships between scenarios and patterns of facilities

scenarios / patterns of facilities	reference scenario	growth scenario	shrinkage scenario
1. maintaining present course			
2. top care			
3. towards a reciprocal aid society			

When the various patterns of (health) care facilities are viewed against the background of the various contexts, the weak points of the scenario method also come to light. They are not forecasts precisely because the future developments involved do not lend themselves to the making of reliable and precise forecasts. In general a test situation which has been designed for that purpose cannot be considered as a forecast. At best it may be considered a 'prediction' of the most likely extreme situations under which the given facilities must operate and meet requirements. A second limitation of the scenario method is that to make situations sufficiently clear and wieldable for the making of policy choices, it is often necessary to present both the situations and the policy alternatives in a highly simplified manner. As a result of this simplification the user is often confronted with questions which fall just outside the limits of the scenario report in question. It is then left to the user to solve the problem by means of supplementing the background study, the contexts, the disturbing developments and the patterns of facilities.

1.4. Design of the report

Chapter 2 deals with the reference scenario. This involves a projection of developments from 1984 and preceding years. When making this projection, it was found necessary to differentiate according to various variants, namely a null variant, a trend variant and a policy variant.

Chapter 3 outlines the medical and medical-technological developments expected. Information deriving from the literature, from experts interviewed and from discussion in the scenario committee on ageing is assembled in this chapter. A synopsis of the group discussions with medical and medical-technological experts is given in appendix I, as well as a list of the participants in the group discussions.

Chapter 4 deals with the growth scenario. By comparison with the reference scenario, this scenario sketches an even greater increase in the demand for (health) care facilities.

Chapter 5 presents the shrinkage scenario as antithesis of the growth scenario. This scenario proceeds from a decreasing (growth in the) demand for facilities.

Chapter 6 supplements the other scenarios by devoting attention to disturbing developments. Two situations are chosen which, should they come about, would exert considerable influence on the whole system of care: a postponement of dementia by 5 years, and a lessening of intergenerational solidarity with respect to the elderly.

Chapter 7 concludes the report with a discussion of the ways in which the results of the scenario project could be employed. The patterns of health care facilities for the elderly are further scrutinized. And finally, the interrelationships between contexts and patterns of facilities are assessed.

Notes chapter 1

(1) For instance De Leefsituatie van de Nederlandse bevolking van 55 jaar en ouder 1976 (Living Conditions of the Dutch Population of 55 years and older 1976). Central Bureau of Statistics, The Hague, 1977, 1978, 1979, 1981 and 1983.

(2) The Elderly and their Health in the Netherlands 1984-2000: Towards a Scenario Report. Utrecht, 1984.

(3) Department of Planning and Policy, 1981.

(4) Beck, 1983 (including SHELL's experiences with scenarios).

(5) In the column 'relative' the absolute demand for facilities is related to the total number of persons of 65 years of age and older. Units such as 'number of hours of care per person' and 'number of days in a nursing home per patient' etc. thus come about. The symbols +, 0 and - are used to indicate an increase, constant demand, or decrease of these units in the year 2000 by comparison with the present situation. See also note 59 in chapter 2.

(6) + = an increase by comparison with the present situation
0 = no change in demand
- = decrease by comparison with the present situation
Further shades of meaning will be attached to the signs + and - (strong, weak + or -) in scenarios B and C. In scenario A figures are given to support the use of the signs.

2 Scenario A: The reference scenario

2.1 Introduction

In this chapter the 'Reference Scenario Ageing' is discussed.
A reference scenario serves as a background against which one or more other scenarios can be viewed. A reference scenario can assume the form of a **trend scenario**. In this instance, existing development trends are extrapolated into the future. A reference scenario can also consist of a **null scenario**. In this case, one or more components are assumed to remain permanently or temporarily unchanged.[1)]
In the 'Reference Scenario Ageing' a combination of a trend and a null scenario have been employed.
In the reference scenario attention will be devoted to the following aspects:

(1) demographic development;
(2) the state of health of the elderly (in relation to medical-technological aspects);
(3) social developments, as far as these are relevant to ageing;
(4) demand for facilities by the elderly;
(5) the economic context.

The reader is referred to Table 1.1 in chapter 1 for an overview of this scenario. This perhaps calls for some explanation. The demographic development discussed in Section 2.2 may be viewed as a trend development (incidentally this demographic development also applies with respect to scenarios B and C). The state of health of the elderly (in relation to medical-technological aspects) is in this scenario assumed to remain unchanged (Section 2.3). The same applies to certain social developments (behaviour and attitudes with respect to health, work, retirement and intragenerational solidarity). These components are not discussed further in this scenario. The social components which in this scenario are assumed to alter by comparison with the present situation are discussed in Section 2.4.

As regards the demand for facilities, it will be necessary to proceed form two variants, namely a null and a trend variant. In the **null variant** the future demand for facilities is calculated by applying present usage figures (sex and age specific) to future demographic development. The **trend variant** proceeds from development of usage in the past (partly the consequence of policy). As will be seen in Section 2.5, these two calculation methods yield considerable differences. These two variants will be viewed against the background of policy proposals with respect to facilities. In this so-called **policy variant**, projections based on norms with respect to capacity as these apply for the various facilities, are presented. By this means, discrepancies between policy resolves and estimates according to the null and the trend variant are made clear.

Finally, the economic context shows how national income should develop if the absolute growth of facilities for the elderly brought about by demographic developments (null variant) is to be met.

The reference scenario thus contains elements which are assumed to remain unchanged and elements whereby existing trends are extrapolated into the future.

2.2 Demographic developments

2.2.1 Population size

In 1983 the Netherlands had nearly 14.5 million inhabitants, of whom 3 million were older than 55 (21.3%) and 1.7 million older than 65 (11.8%). In coming years the population of the Netherlands will age even further. The share of the elderly in the population continues to grow on account of the diminishing share of the youth brought about by the sharp decline in fertility.

For the purpose of forecasting future population size, the medium variant of the Central Bureau of Statistics population forecast dating from 1982 has been employed.[2)] The medium variant was chosen to make possible a comparison with the estimates made by the Social and Cultural Planning Bureau with regard to the trend variant. The difference between the medium and low variants with respect to the age group 55 years of age and older is minimal. Consequently there are hardly any differences in the calculations if the low variant is employed, except in cases involving the relative share of the elderly in the total population.

Table 2.1 shows the growth of the category 55 years of age and older.
Not only is there an absolute growth of the category of 55-plussers, but the relative share of this category in the total population increases from 21.3% in 1984 to 23.4% in 2000 (65-plussers: from 11.8% in 1984 to 13.4% in 2000).
Indicentally, compared with other European countries this latter percentage is not very high. In a number of countries the share of 65-plussers expected in the Netherlands in the year 2000 has already been exceeded.[3)]

Table 2.1 Number of elderly from 1984-2000, absolute (x 1000) and in percentages of the total population

	1984	1985	1986	1987	1990	2000
Total population	14.473	14.560	14.646	14.730	14.973	15.643
55 - 59	704	704	713	718	729	857
60 - 64	664	681	676	672	668	717
65 - 69	538	533	555	572	626	632
70 - 74	463	469	471	475	464	534
75 - 79	348	354	358	360	376	437
80 - 84	215	222	229	234	248	261
85 +	146	151	156	162	181	225
% 55+ in tot. pop.	21.3	21.5	21.6	21.7	22.0	23.4
% 65+ in tot. pop.	11.8	11.9	12.1	12.2	12.7	13.4
% 75+ in tot. pop.	4.9	5.0	5.1	5.1	5.4	5.9
% 85+ in tot. pop.	1.0	1.0	1.1	1.1	1.2	1.4

Source: Derived from the CBS forecast 1980, medium variant. For more detailed data see Appendices A and B.

The population will continue to grow till 2010 - according to the medium variant - after which it will decrease gradually (see Appendix A). Data regarding long-term development of the population as a whole are however somewhat speculative as a consequence of the uncertainties attaching to fertility and migration. In 2030 approximately 5.4 million people will accrue to the category of 55-plussers and 3.2 million to the category of 65-plussers (21.2%). These figures show that not only the absolute number of elderly, but also the relative share in the total population increases.[4)]

This is not the only factor which determines the problem of ageing. This problem is determined by the total demographic burden (0-19 year-olds and 65-plussers) in relation to the potential working population (20-64 year olds), by the share of the labour force that is also dependent on collective resources (the unemployed, disabled persons, people who have retired prematurely, etc.), and by the level of the costs of facilities for the elderly.

2.2.2 Composition of the category of elderly

Three developments characterize the present and future composition of the category of elderly, namely a large surplus of women, an increasing number of persons living alone and an increasing number of very old persons (80 years of age and older).[5)]
These developments are closely related, since the higher average life expectancy of women means that a lot of women survive their husbands. Consequently the category of persons living alone contains more women than men. The same applies to the category of the very elderly, which is composed mainly of women.

Appendix B gives an overview of the development of the absolute number of elderly according to sex and civil status till the year 2030 (medium variant). Cf. also Figure II.1.

In the period 1980-2030 the number of men in the age category 55 years of age and older who live alone will increase from 249,000 to 982,000 and the number of women living alone (unmarried, widowed and divorced) from 765,000 to 1,695,000.

The data relate to the Netherlands as a whole. Large differences are however to be found within the Netherlands. The Background Study contains data regarding the regional distribution of the older population.

Figure II.1 Population according to age category and sex, 1st January 1980 and 2000

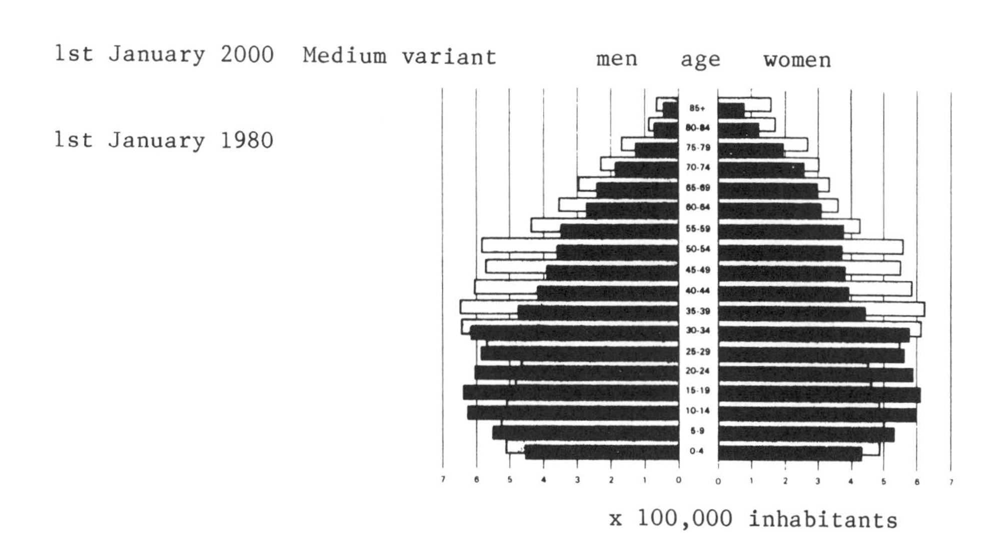

Source: CBS forecast of the population of the Netherlands after 1980, p. 43.

2.3 The health status of the elderly

To give a brief sketch of the state of health of the elderly, one must be selective. Four indicators have been chosen for describing the state of health of the elderly, namely:

(1) average life expectancy and maximum duration of life;
(2) mortality according to cause;
(3) morbidity (disease, diverse complaints) and invalidity;[6]
(4) subjective health.

(1) Average life expectancy and maximum duration of life
Table 2.2 gives data regarding life expectancy at birth in the past.

Table 2.2 Life expectancy at birth according to sex, 1966-1982

	1966-1970	1971-1975	1976-1980	1981	1982
men	71.0	71.2	72.0	72.7	72.8
women	76.4	77.2	78.6	79.3	79.5

Source: CBS

As this table shows, life expectancy is rising both for men and for women. There is however a difference in life expectancy between the sexes. Various explanations are given for this, such as differences in lifestyle (smoking, consumption of alcohol), different conditions of work, culturally determined sex roles, and biological differences.

Many uncertainties attach to the question of how this gap in life expectancy between the sexes will develop in the future. For instance, the assumption that the gap will be reduced by emancipation processes is not borne out by the situation in the Soviet Union. Of all European countries, the difference in life expectancy is largest in that country, namely 11,6 years. How smoking behaviour will affect the gap is also unclear. This question warrants some further discussion.

American studies[7)] have shown that there has been a levelling off in the incidence of lung cancer among men in the last decade, while the incidence among women shows a sharper increase than among men, without any signs of levelling off. Taking it that there is a relationship between smoking and lung cancer with a latent period of 20 to 30 years, this tendency among women could be explained i.a. by the increasing consumption of cigarettes among women in the sixties and early seventies. It however remains an open question in how far this will serve to reduce the gap in life expectancy between men and women. There are a number of arguments against this assumption:

- The incidence of lung cancer among women remains markedly lower than among men despite a possible increase among women in the past (cf. Figure II.2).

Figure II.2 Estimated incidence of lung cancer in 1980 in the Netherlands according to sex and age

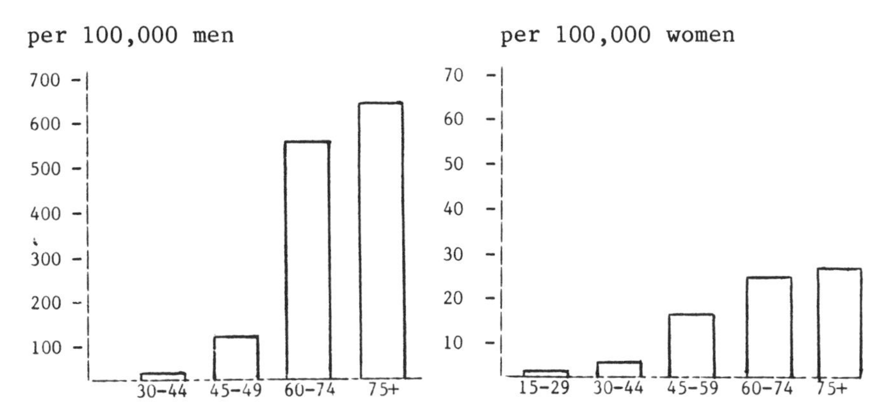

Source: Scenario Commission on Oncology

- There is a growing tendency to reduce the quantity of tar in cigarettes and to consumption of more filter cigarettes. The effect of this will probably be that the incidence of lung cancer among women will not reach the present level among men, since women will benefit more from these shifts.

- After an initial rise in the percentage of smokers among women in the sixties and early seventies, a decrease is again observable since 1975, as borne out by Table 2.3. As opposed to this however, there has been a per capita increase in the number of cigarettes smoked, the increase being greater among women than among men.
  The most recent figures from the Foundation for Health and Smoking[8)] for 1983 show that the percentage of smokers has again risen slightly among both sexes. The percentage of smokers among women however remains lower than among men.

For the time being, with respect to scenarios A and B, it is assumed that there will be hardly any narrowing of the gap in life expectancy between men and women. This has important consequences for the aid with which partners provide each other. In 1982 the difference in life expectancy at birth between men and women was 6.7 years (cf. Table 2.2). In accordance with the CBS forecast employed, a slight decline in the difference in life expectancy between men and women is assumed for scenarios A and B. This decline is actually brought about by an increase in the life expectancy of men to 73.3 years. After 1990 no further changes in life expectancy are foreseen in the forecast.[9)]

Table 2.3 Percentage of smokers according to sex, 1958-1982

	1958	1963	1967	1970	1975	1979	1981	1982	1983
M	90	82	78	75	66	52	47	41	44
F	29	32	42	42	40	38	36	33	35

Source: Foundation for Health and Smoking

With respect to scenario C however, long-term (after 2000) a further decrease in the difference in life expectancy between the sexes is assumed. This is consistent with the assumption from which the scenario proceeds that the lifestyles of men and women will become more similar.

In general 110 to 115 years is assumed to be the maximum age which man can attain. No changes with respect to this are expected in the future, since it is assumed that longevity technology will not be capable of significantly extending duration of life before the year 2000.

(2) Mortality as to cause
After the 'epidemiological transition' (i.a. the disappearance of infection and deficiency diseases as primary causes of death, which were replaced by prolonged, irreversible diseases) there is no sign of a stable situation with respect to mortality and morbidity. Mortality and morbidity patterns are still undergoing large changes, also among the elderly. Ex post, it is sometimes possible to find reasonable explanations for developments, but reasonably acceptable projections are still in the design stage (for instance other scenarios). If all the separate uncertainties are taken together, it must be concluded that at present the predictability of the state of health of the population is fairly limited.[10)]

Mortality among the elderly results mainly from coronary and arterial diseases and cancer, men and women showing different patterns of mortality. In the reference scenario it is assumed that coronary and arterial diseases and cancer will continue to be the most important causes of death till the year 2000 (see Appendix D). As a consequence of ageing of the population, there will naturally be an increase in the absolute number of deaths per annum.[11)] Moreover, determining the primary cause of death will be complicated by multiple pathology among the elderly. Some diseases, such as dementia and rheumatism, are not considered to be primary causes of death, but certainly exert a negative influence on life expectancy.

It would appear likely that the system of registration of cause of death will be improved before the year 2000. It is expected that before this date euthanasia (cf. also Section 2.4.6) will also be listed as a cause of death.[12)] Registration will also become more reliable as a consequence of better diagnosis and performing of postmortems in unclear cases.

(3) Morbidity

From a point of view of health care (facilities), morbidity data relating to the older population are more important. Data with respect to diseases and invalidity encountered among the elderly are however inadequate. The morbidity data required are not those which show diseases peculiar to the elderly, but those covering all diseases encountered among the elderly. A distinction can be made in these data between (a) diseases resulting specifically from the process of ageing (deterioration of the function of an organ - system; initially compensation mechanisms can deal with this deterioration, but eventually the stage is reached where symptoms must be classified as pathological); and (b) diseases with a long incubation period; some of these diseases develop at an early stage, but only become manifest at an older age.

It is characteristic that a large percentage of diseases among the elderly must be classified as chronic-degenerative diseases which are seldom or never classified as causes of death, but are certainly responsible for a high degree of dependency and the need for care (for instance chronic afflictions of the joints).

In as far as morbidity data are available, in general these yield no insight into the combination of afflictions in one patient.

Precisely this so-called multimorbidity is characteristic of (very) old patients. [14)]

We restrict ourselves mainly to the survey 'The Living Conditions of the Population of 55 Years of Age and Older' (1982) carried out by the CBS[15)] for data relating to morbidity among the elderly. In this survey a distinction is made between elderly persons living independently and those residing in homes for the elderly.[16)]

Appendix C gives an overview of independently residing elderly people who suffer from afflictions, according to the nature of the affliction. Percentages show that there is a certain degree of multimorbidity. The most common complaints are 'rheumatism' (among 33% of the inverviewees of 55 years of age and older), 'shortness of breath' (21%), 'high blood pressure' (20%) and 'dizziness' (15%). Interviewees were also asked if they suffered from other complaints. Answers to this question were however often difficult to interpret (for instance complaints after an operation).

Of the people of 55 years of age and older living independently, by far the majority could hear and see well, even though, as might be expected, these senses deteriorated with age.
Among the inhabitants of homes for the elderly the situation was less favourable with respect to these senses (sight is an important criterion for admission to a home for the elderly). It should however be pointed out that possibilities for correcting deficiencies of sight are in general greater than for correcting deficiencies in hearing.
Appendix E gives an overview of the (prevalence of) disease among the elderly.

In how far afflictions lead to invalidity is not unequivocally evident. For instance, the diagnosis groups 'rheumatic complaints' (especially women) and 'coronary and arterial diseases' account for a large percentage of handicapped persons. The type of functional disturbances most often encountered are disturbances in stamina, in walking, in function of arm and hand (especially among women) and in balance.[17)]

The state of health, measured on the ADL scale[18)] yields the following picture: 77% of all independently residing elderly people can perform all activities without difficulty and without aid (55 years and older). Among residents of homes for the elderly this percentage is markedly lower, namely only 23% (see Appendix F). Compared with the results of the living conditions survey of 55-plussers carried out in 1976 (see Appendix F), there would appear to be a worsening of the state of health of inhabitants of homes for the elderly (in 1976: 43% of all activities could be performed without difficulty or aid). This tallies with expectations in view of the more stringent indication and admission policy adhered to since 1976.

The state of health of independently residing elderly people could also be affected by this admission policy. Indeed it is found that the category who can perform 'all activities without difficulty' has dropped from 83.2% to 77%. When we add to this the categories 'one activity with difficulty and all without aid' and 'two or more activities with difficulty and all without aid', the percentages are found to be 93.8 (in 1976) as against 95 (in 1982). The percentage of independently residing elderly people who require no assistance in ADL has thus remained fairly constant.

Four percent of independently residing elderly people older than 55 require assistance in ADL (1% unknown). In view of the fact that the need for assistance increases with age, and that women require more assistance than men, future demographic developments imply a significant absolute increase in the number of persons who will require assistance. Relatively speaking, however, the reference scenario assumes the number of people requiring assistance in ADL to remain constant.

Morbidity data relating to psychic disturbances are even scarcer than with respect to somatic complaints. For instance there are hardly any epidemiological data on the various types of dementia in the Netherlands (except for the elderly in intramural institutions). Prevalence figures from other countries vary from 0.5 to 31.8%, figures also differing for various age groups and sexes. Chapter 6 will discuss dementia in detail. Appendix D also includes data on depression among the elderly.

Double ageing has consequences for the future pattern of morbidity, both quantitatively and qualitatively. To give an impression, we present some data on GP practices. Van den Hoogen et al. have calculated what shifts will take place in 'standard GP practices' as a consequence of the altered composition of the population.[19)]

The conclusion drawn is that in 2000 there will be an annual increase of 5% in the total number of new and recognized afflictions among both men and women (see Table 2.4.a). The increase in new afflictions is not impressive (approximately 2% among men and 1% among women, see Table 2.4.b), but as against this, there will be a considerable increase in the number of already recognized afflictions: 17% among men and 15% among women (see Table 2.4.c).[20]

Table 2.4 Shifts in morbidity, 1984-2000

(a) Complaints encountered among men and women in a 'standard GP practice' (N=2800) per age category in the years 1980 and 2000 respectively (in absolute and index figures). This concerns **all** complaints encountered (exclusive of partus, gravidity, counselling etc.)

age	men-1980		men-2000		women-1980		women-2000	
(years)	(abs)	(index)	(abs)	(index)	(abs)	(index)	(abs)	(index)
0 - 19	833	100	711	85,4	820	100	676	82,4
20 - 59	1480	100	1649	111,4	1914	100	2077	108,5
60 +	621	100	730	117,6	1023	100	1200	117,3
total	2934	100	3090	105,3	3757	100	3953	105,2

(b) Complaints encountered among men and women in a 'standard GP practice' (N=2800) per age category, in the years 1980 and 2000 respectively (in absolute and index figures). This concerns **all new** complaints (exclusive of partus, gravidity, counselling, etc.)

age	men-1980		men-2000		women-1980		women-2000	
(years)	(abs)	(index)	(abs)	(index)	(abs)	(index)	(abs)	(index)
0 - 19	775	100	665	85,5	776	100	641	82,6
20 - 59	1180	100	1284	108,8	1455	100	1545	106,2
60 +	293	100	343	117,1	467	100	539	115,4
total	2248	100	2292	102,0	2698	100	2725	101,0

(c) Complaints encountered among men and women in a 'standard GP practice' (N=2800) per age category, in the years 1980 and 2000 respectively (in absolute and index figures). This concerns **already known** complaints (exclusive of partus, gravidity, counselling, etc.)

age (years)	men-1980 (abs)	men-1980 (index)	men-2000 (abs)	men-2000 (index)	women-1980 (abs)	women-1980 (index)	women-2000 (abs)	women-2000 (index)
0 - 19	51	100	41	80,4	43	100	33	76,7
20 - 59	292	100	355	121,6	449	100	519	115,6
60 +	303	100	357	117,8	546	100	646	118,3
total	646	100	753	116,6	1038	100	1198	115,4

Source: Van den Hoogen et al., 1982, p. 872 (adapted)

The increase in complaints among the people in the age category 60+ varies from 15 to 18%. Here again it should be pointed out that future developments in health are difficult to predict, and that the demand for aid is determined not only by the state of health but also by other equally unpredictable factors. This means that any projections on this subject are of only limited value.

In as far as qualitative changes are concerned, particularly the share of chronic complaints will increase (arthritis deformans, hypertension, ischemic coronary diseases and deafness). See Appendix G on this subject. It would appear that some change is coming about in the ranking order of the twenty most often presenting complaints.

(4) Subjective health

Subjective health, that is to say, the manner in which a person perceives his own health, has been chosen as the final indicator of the state of health of the elderly.
The living conditions survey of 55-plussers carried out in 1982 among independently residing elderly persons of 55 years of age and older, shows that 51% consider themselves to be healthy, 30% reasonably healthy, 13% slightly ailing, and 5% ailing. These percentages are 29, 35 and 26 and 10 respectively among residents of homes for the elderly.[21)] There are variations in percentages per age group and per sex.

One's subjective view of his own health does not always tally with the 'objective' state of health. Ago-specific criteria are often applied ('old age comes with infirmities'), while the residential situation may also influence the outlook ('I live in a home for the elderly, so naturally I can't be very healthy').

What conclusions may be drawn from the above with respect to the state of health of the older population till the year 2000?
The 'average' state of health of the population of 55 years of age and older will deteriorate exclusively as a consequence of demographic developments (double ageing), taking it that all other conditions remain the same (lifestyle, medical-technological possibilities, no spontaneous changes in the increase or decrease of certain diseases, etc.). The increase of half a year in the average life expectancy at birth for men which is anticipated till the year 2000 may at the individual level result in a longer period of illness before death. This would not be the case among women since the average life expectancy of women at birth is assumed to remain constant.

Both these developments lead to a quantitative increase in the total number of complaints and a qualitative shift in the pattern of complaints.

2.4 Social developments

2.4.1 Introduction

In this section a number of social developments are discussed which could be of importance for the position of the elderly and the care which they receive. The issues involved - as opposed to socio-economic issues which vary considerably according to the economic situation of the country - are developments which show a clearly continuous trend. These developments are:
(1) an increasingly higher level of education among the elderly;
(2) increasing emancipation of the elderly;
(3) reduced number of children, i.a. of importance in connection with (a decrease in) umbrella care of their parents by children;

(4) increasing emancipation of women, of importance in connection with the decreasing possibilities for women to provide umbrella care;
(5) increasing tolerance of (active and passive) euthanasia;
(6) improved pensions.

The social trends presented are assumed to be relevant not only to the reference scenario, but also to scenarios B and C. That is to say: the **direction** of the trends presented (+ or -) is the same for all scenarios, though the **intensity** of change can vary per scenario for some trends.

2.4.2 Educational level of the elderly

The educational level of the elderly will rise. In general, younger generations have received more schooling than the older generations. This applies both for men and for women. This is illustrated in Table 2.5. This table shows the highest level of education achieved for which a certificate of any kind was received.

Table 2.5 derives from the living conditions survey of 1980. The 35-44 year olds and the 45-54 year olds from this survey constitute the elderly of the year 2000 (55-plussers). Their educational level - as shown by Table 2.5 - is clearly higher than that of the present elderly.

What consequences may be expected from this increasingly higher level of education? Havighurst[22] mentions among others the following: greater activity, greater concentration on achievement, greater satisfaction with life, less personal problems, greater participation in politics and volunteer work. In general it may be said that older people with a higher level of eductional behave 'younger' than older people with a lower level of education. Various Swedish studies[23] have however shown that people with a higher level of education are more inclined to make use of health care facilities than people with a lower educational level. One reason for this might be that people with a higher level of education are better informed as to the available health care facilities and access to these facilities.

Table 2.5 Educational level according to age and sex in % (in 1980)

	only primary	lower voc. train.	sec.mod.	grammar school	higher voc.tr./ univ.	abs. (=100%)
MEN						
18-24	10	23	21	40	6	391
25-34	13	21	6	39	20	779
35-44	17	20	4	39	20	589
45-54	26	22	6	32	14	475
55-64	33	22	5	28	13	385
65-74	48	19	5	21	6	252
75 a.o.	62	15	3	11	8	132
	--	--	--	--	--	----
total	23	21	7	34	15	3003
WOMEN						
18-24	11	20	21	44	5	403
25-34	18	26	14	27	15	664
35-44	27	28	13	24	9	499
45-54	43	20	11	17	9	413
55-64	53	21	8	12	6	332
65-74	71	9	5	9	6	267
75 a.o.	77	6	3	10	5	145
	--	--	--	--	--	----
total	35	21	12	23	9	2723
MEN + WOMEN						
18-24	10	21	21	42	5	794
25-34	15	23	10	34	18	1443
35-44	21	24	8	32	15	1088
45-54	34	21	8	25	12	888
55-64	42	22	6	20	10	717
65-74	60	14	5	15	6	519
75 a.o.	70	10	3	11	7	277
	--	--	--	--	--	----
total	29	21	10	29	12	5726

Source: CBS, 1984, p. 43.

Older people with a higher educational level will also be more inclined to participate in adult education and educational courses for the elderly. It is not implausible that, as in France and Belgium, in the Netherlands too a counterpart of the "Université du Troisième Age" will come into being, in which the elderly will participate both as instructors and as students. In addition the elderly will also take an increasing interest in other new forms of education (Open University, Open School, Multimedia education, Summer schools, etc.).

2.4.3 Emancipation of the elderly

Another development which is closely connected with the above, is the expected increasing emancipation of the elderly in society. Emancipation of the elderly focuses on furthering equality of the elderly in society. The elderly must not (or no longer) be viewed as encumbrances requiring only care and guidance, but rather they should be accorded an important say in determining the circumstances under which they live.[24)]

The emancipation of the elderly is one of the goals of the organizations of the elderly, united in COSBO Netherlands (The Central Organ of Co-operation Unions of the Elderly).[25)] COSBO Netherlands co-operates at a national level. In addition to this, co-operation also takes place between the unions at a local and provincial level. For instance, these unions hold over 30% of the management seats of the projects Co-ordinated Work for the Elderly (GBW).

It is anticipated that in the future more elderly persons will join these unions than might be expected purely on the basis of demographic developments. The group of unionists will become increasingly assertive, partly as a result of the better educational opportunities of the last decade. More persons will be trained for positions of leadership. Demands to government[26)] will be more explicitly formulated, for instance the installation of a Council for Policy with Respect to the Elderly, representation by the elderly in important councils (Social Economic Council, Health Insurance Fund Council), legal regulation of democratization and methods of dealing with complaints in nursing homes and homes for the elderly, the introduction of flexible retirement, inclusion of the age criterion in antidiscrimination legislation, the continued linkage of old age pensions to the minimum wage (which should no longer be reduceable), etc. As far as efforts to assist the elderly to continue to live independently in their own environment are concerned, the policy of unions of the elderly at the moment runs parallel with that of government.

For the time being, it would not appear likely that the formation of a political party of elderly persons will be one of the means employed for furthering the interests of the elderly. It may however be assumed that there will be no (sudden) change in the voting behaviour of the elderly by comparison with their voting behaviour at an earlier stage.[27)] Rather, questions relating specifically to the elderly will be brought to the attention of the public by systematic lobbying of the unions of the elderly and by means of ad hoc actions.

2.4.4 Reduced number of children

As is known, in the past decade there has been a reduction in fertility. The population forecast which we employ[28)] proceeds from the assumption that the average number of children per woman will drop from approximately 2 for the birth generation of around 1945 to slightly more than 1.6 (low variant of the population forecast) or slightly more than 1.8 (high variant) for women who were born around 1970 (Figure II.3).

For our purposes, we are however interested in the cohorts which in the year 2000 will accrue to the category of the elderly (55-plussers). These are the birth cohorts of 1945 and earlier. The fertility level of these cohorts is already more or less known, since the majority of women from these cohorts have passed the reproductive phase. As is shown in Figure II.3, the fertility level has declined considerably for most of these cohorts: whereas the average number of children per woman of the birth generation 1930 was approximately 2.6, the average number of children for the birth generation 1945 is approximately 2.

Figure II.3 Average (expected) number of children per woman (birth cohorts 1930-1980)

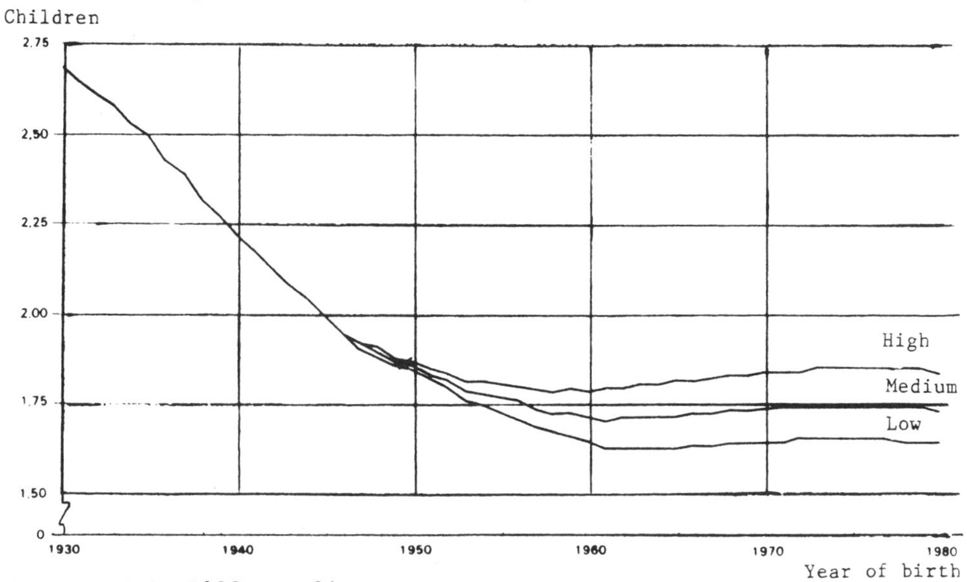

Source: CBS, 1982, p. 34.
On account of the small differences between low, medium and high variant, for the average number of children per woman of the birth cohorts 1930 to 1947 only the medium variant has been traced.

The reduction in the size of families can prove to be an important factor, since it could influence the amount of umbrella care with which children can provide their parents. Moreover, the woman is younger when her last child is born, which means that this child would already have left some several years before the parents require assistance. In addition to this there are other factors which might exert a negative influence on umbrella care (increasing individualization, increasing geographical mobility, increasing emancipation of women, etc.).

There is another phenomenon which warrants attention in this connection, namely voluntary childlessness. It may be expected that both voluntary childlessness and the social acceptance of this phenomenon will increase. Figure II.4 gives information on this topic.

Figure II.4 Percentage of lastingly childless marriages per marriage generation

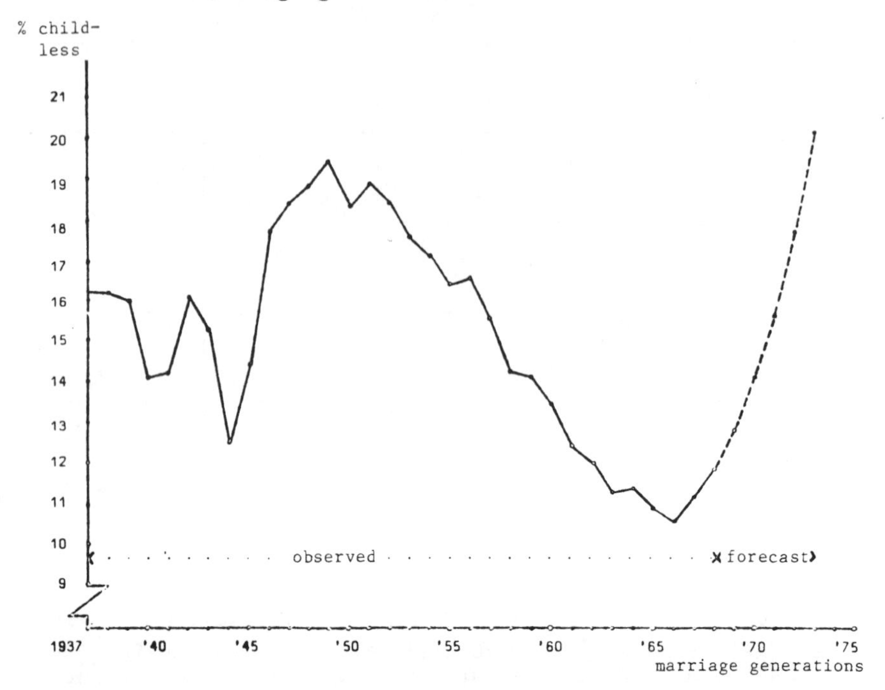

Source: Frinking, 1975 and Niphuis-Nell, 1979.

The dotted line on the right side of Figure II.4 indicates the expected future percentage of lastingly childless marriages. Taking it that of the total percentage of childless marriages 7 to 8% are involuntarily so, then for the younger marriage cohorts a rise in the percentage of voluntarily childless marriages from 3% for the 1966 cohort to 13% for the 1973 is observable. What the consequences of this childlessness will be for the care of the elderly must be viewed from the perspective of three (or more) generations. These generations include the childless elderly who cannot count on any umbrella care from children.
Assuming an average marriage age of 25 and that the greatest need for umbrella care among the elderly will be when they are very old, for the purpose of assessing umbrella care around the year 2000, attention focusses on marriage cohorts dating from around 1950. As is shown in Figure II.4 there was a peak in childlessness around 1950. In this connection an estimate of the Social and Cultural Planning Bureau is also relevant, namely that in the year 2000 the

percentage of couples that (voluntarily) remain childless will be around 25%.[29)]
On the other hand, there will be the childless middle-aged people who will be very important for the umbrella care of their own parents, since it may be assumed that childless persons of middle age will be one of the first sources of assistance approached when their elderly parents require assistance. With respect to the year 2000 the marriage generations shown in the extreme right of Figure II.4 are relevant.

Summarizing, it may be concluded that childlessness will thus exert both a negative and a positive influence on possibilities for umbrella care.

2.4.5 Emancipation of women

The process of emancipation of women will be illustrated below on the basis of the indicators: (1) participation in education and (2) participation in the labour process.

As far as participation in education is concerned it may be posed that women to a large degree catch up with the former educational lag (see also Table 2.5 for sexes and age categories separately). Girls participate increasingly longer in education, and adult women show a growing interest for courses of education.

Despite increasing unemployment, the participation of (married) women in the labour process has risen in recent years: in 1977 26% of married women performed paid work, in 1979 28% and in 1981 33%.[30)] It is expected that in the future these participation percentages will increase, especially if there should be a shortening of working hours (see scenario C).

Since the sixties there has also been an increase in social tolerance of the working married women, and at the moment a stabilization of this phenomenon is to be observed, as shown in Table 2.6.

It may be expected that as a result of the above-mentioned developments, the willingness and possibilities for women to provide informal aid will decrease (for instance aid relationships between daughters (in law) and parents (in law) who require assistance.[31])

Table 2.6 Attitudes with respect to participation of married women in labour process in percentages

(a) WORKING IS

	to be recommended	no problem	undesirable	abs.
1965	1.7	14.6	83.7	1503
1970	16.1	40.3	43.6	1880
1975	13.6	44.1	42.3	1652
1978	15.9	40.9	43.2	880
1979	11.4	54.6	33.9	1759
1980	12.7	51.9	35.5	1699
1981	11.1	51.2	37.7	1721
1983	13.2	51.0	35.8	1705

(b) PAID WORK SHOULD BE DIVIDED EQUALLY BETWEEN MEN AND WOMEN

	strongly in favour	in favour	partly for, partly against	strongly against	abs.
1981	21.8	35.1	17.1	26.0	1885
1983	25.3	32.4	16.7	25.5	1807

(c) DOMESTIC WORK SHOULD BE DIVIDED EQUALLY BETWEEN MEN AND WOMEN

	strongly in favour	in favour	partly for, partly against	strongly against	abs.
1981	24.7	36.1	17.5	21.7	1935
1983	28.0	35.5	17.2	19.4	1841

Quoted in SCP, 1984, pp. 303 and 304.

2.4.6 Tolerance of euthanisia

There has been a significant increase in tolerance of euthanasia since the sixties. In the various studies on this subject however, figures as to proponents and opponents of euthanasia are extremely diverging. This is possibly a consequence of the manner in which questions are posed (for instance whether or not a distinction is made between active and passive euthanasia). Survey results with respect to attitudes of the Dutch population are given in Table 2.7.

Table 2.7 Opinions on euthanasia in 1966-1984 (in percentages)

	1966	1975	1979	1980	1981	1983	1984
(a) possibility of euthanasia							
- should be permitted	39.9	52.6	51.4	52.4	53.8	56.4	
- depends	11.5	23.7	32.4	35.5	32.8	31.2	
- should not be permitted	48.6	23.7	16.0	12.1	13.4	12.5	
(b)							
- agreement with passive euthanasia			19.0				13.0
- also agreement with active euthanasia			56.0				77.0
- disapproval of euthanasia			23.0				)
							) 10.0
- no comment			2.0				)

Source: Survey of Religion in the Netherlands, 1966.
Survey of Progressionism and Conservatism, 1970.
Cultural changes in the Netherlands, 1958-1975, 1979, 1980, 1981, 1983.
Lagendijk Opinion Survey, 1979, 1984.

As yet, little is known regarding the attitudes of the elderly themselves to euthanasia. Hilhorst and Verhoef reported on a survey in two nursing homes regarding guidance to the dying and euthanasia.[32] It was found that people had difficulty discussing the subject. At the end of the seventies the Conditional Request for Euthanasia was still a considerable exception in nursing homes.

Data as to the degree to which patients request euthanasia are scarce and inadequate (partly to be attributed to the punishability of euthanasia). Since 1976 however the gauge stations of the Continuous Morbidity Registration have kept check of requests for active euthanasia (either direct or indirect[33]. Table 2.8 gives an overview of these figures.

Table 2.8 Absolute number of patients who requested active euthanasia of the GP according to age 1976-1982

	<55	55-64	65-74	75-84	≤ 85	totaal
1976	2	4	3	3	3	15
1977	2	3	2	2	-	9
1978	3	2	3	2	-	10
1979	3	7	12	2	4	28
1980	2	5	5	7	3	22
1981	8	4	5	10	3	30
1982	-	6	10	8	2	26

Source: Continuous Morbidity Registration Gauge Stations, 1982.

In recent years the absolute number of requests for euthanasia has been found to remain fairly constant. Relatively speaking, patients with a malignancy are strongly represented in these requests.
As far as the relative number of requests for active euthanasia is concerned, no pronouncements are possible since no data are available concerning the number of persons whose conditions are such that they might be inclined to request euthanasia (morbidity data).[34]

Recently some data have been published as to the attitude of GPs to euthanasia. Table 2.9 gives an overview of these data.

Actually, the distinction between the various forms of euthanasia (passive, active, direct, indirect) was recently rejected by the Royal Dutch Society of Medicine. In their opinion this distinction would confuse the issue of euthanasia.

Table 2.9 Opinions of GPs with respect to euthanasia 1984

-	proponent of active euthanasia	37%
-	opponent of active euthanasia	31%
-	objects to expressing opinion in set terms in view of complexity of the problem	31%
-	proponent of passive euthanasia	83%
-	opponent of passive euthanasia	4%
-	no comment	12%
-	in favour of relaxing legislation	61%
-	favours maintaining present legislation	30%

Source: Intomart.

For the scenarios, we expect that the trend towards increasing tolerance of euthanasia will continue. Present regulations governing punishability will be relaxed. It seems not unlikely that the conditions under which euthanasia is acceptable will be laid down in a new law, which will also draw up regulations for the supervision of euthanasia. It will then become possible to quote euthanasia officially as a (contributory) cause of death.
The increasing tolerance of euthanasia could influence the quality of the (end of) life of the elderly. However no significant changes in the quantity of care of the elderly may be expected from such a shift.

2.4.7 Improvement of pensions

The first pension regulation in the Netherlands came into being in the middle of the nineteenth century. However, the most important expansion of the pension regulations took place in the nineteen fifties and sixties. On the basis of a pension accretion period of 35 to 40 years, this means that the effects of the growth of pension funds will certainly be felt till the year 2000. We will not discuss the various pension systems. Rather, we focus on the range and level of pensions in the past and the future.

As Table 2.10 shows, in the past there has been a growth of the number of people eligible for pensions.

Table 2.10 Number of people eligible for a pension as a percentage of the relevant category of elderly 1964-1981

	COMPANY PENSION SCHEMES		CATEGORIAL PENSION SCHEMES	
year	retirement pensions	widow's pensions	retirement pensions	widow's pensions
1964	6.9	5.4	23.9	10.7
1968	8.5	6.8	30.3	17.2
1972	9.4	8.1	35.2	22.0
1976	10.5	9.4	39.4	27.4
1980	12.0	10.6	44.0	31.6
1981	12.5	10.8	45.2	32.7

Source: Pommer and Wiebrens, 1984, p. 25.

The number of people eligible for a pension as a percentage of the relevant category of elderly is found to have risen in the past years. In the case of retirement pensions, the relevant category of elderly is defined as all men of 65 years of age and older (irrespective of civil status) plus single women of 65 years of age and older. In the case of widow's pensions, the relevant category of elderly consists of all widows of 65 years of age and older. Table 2.10 however does not show the number of elderly persons who receive a pension since, by changing jobs, a person may be the recipient of more than one pension. The percentage of elderly persons with a pension is thus probably lower than shown in Table 2.10. Nonetheless, it may be inferred that the number of recipients of pensions is on the increase.

The level of pensions has also risen over the years as a consequence of the fact that the number of years of service covered by the pension has increased.

As far as the future till the year 2000 is concerned, it is expected that the trend towards increase in the number of recipients and level of pensions will continue.
As far as the elderly in the next century are concerned, however, it is possible that the introduction of drastic shorting of working hours with proportionate reduction of wages or a situation of lengthy unemployment will have the result that there will be a reduction in pensions by comparison with former generations. However, as a consequence of increased participation in the labour process by (married) women, this will not necessarily mean a reduction of the family or household pension. As far as the consequences of lengthy unemployment for the accretion of pensions is concerned, it is expected that measures will be speedily taken to mitigate these consequences.

Even though it would appear that retirement pensions can be maintained more or less stable till the year 2000, considerable uncertainty attaches to maintaining old age pensions in the present form. As opposed to retirement pensions, old age pensions are financed by the ratio system. This system implies solidarity between the generations, since there is a transfer of incomes between the working population at a certain moment and the persons eligible for old age pension at that moment. In 1980 the ratio between the payers of premiums and persons eligible for old age pension was 3:1. In 2030 the ratio will be 1.6:1.
This drastic shift in ratios has recently evoked a large number of publications as to the capacity for meeting payments for old age pensions in the future, and various suggestions were made for solving the question (changing the system of levying of premiums, changes in pensions, changes in the eligibility age for pensions, etc.).

How old age pensions will develop in the future of course to a large degree depends on the economic situation of the moment. This is dealt with in the relevant sections in the three scenarios. Any significant deterioration in old age pensions would however not seem likely, partly in view of the fact that the elderly, united in unions for the elderly, will become a political power of increasing account. The income position of the elderly (retirement and old age pensions) is of importance for the type of facilities of which they make use and the degree to which they use them. This will be dealt with in the respective scenarios.

2.4.8 The position of the elderly in society

Social developments as discussed above influence the position of the elderly and the care of the elderly. A heightened educational level and increasing emancipation of the elderly, as well as the improvement of incomes, will not automatically result in an improved position for the elderly in society. This will depend to a large degree on the general perception of the place of the elderly in society.
In the course of time various theories, based on two points of view, have been formulated. These points of view are: the elderly may no longer and the elderly need no longer participate in society. The various theories are termed: 'expulsion theory', 'activities theory', 'disengagement theory' and a variant of this, namely the 'engagement at a distance theory'.[35] These theories have been subject to much criticism, since two elements always play a role: actual behaviour of the elderly as well as norms with respect to this behaviour. Moreover, insufficient distinction is made as to the points of view of each of the parties involved: government, the providers of aid, younger age categories, and not in the last place, the elderly themselves. It will be clear that various points of view can exist at the same time side by side, even within each of the parties involved. Sometimes the proponents of a certain view formulate this explicitly (for instance in the Memorandum of Policy Relating to the Elderly 1975, in which the authors proposed 'engagement at a distance' as an effective theoretical basis for policy relating to the elderly). In most cases the theoretical points of departure remain implicit. It is not possible to characterize each scenario in terms of one or other of these theories. The scenarios differ with regard to the relationship between the elements of each view among each of the parties concerned. The way in which these views are incorporated into the scenarios will be explained when discussing the various variables.

2.5 The demand for (health) care facilities

2.5.1 Introduction

In this section an attempt will be made to outline developments in the demand by the elderly for (health) care facilities till the year 2000.

As was already said in section 2.1, when making estimates of the future demand for (health) care facilities, it is necessary to proceed from two variants:

(a) **The null variant** In this variant it is assumed that the present sex and age-specific usage will remain constant in the future.

(b) **The trend variant** In this variant account is taken of past developments in sex and age-specific usage of facilities (partly as a result of policy), and these developments are extrapolated into the future.

At the moment existing (health) care facilities may be viewed as resulting from the demand for and the supply of these facilities. In the reference scenario we focus exclusively on **demand** for facilities as far as future developments are concerned, irrespective of whether, in view of policy (with respect to health care) it will be possible to meet this demand. Results are however compared with preferred policy in order to be in a position to indicate discrepancies. This is the so-called:

(c) **Policy variant**

In this estimation method no account is taken of the fact that supply can at the same time modify the demand ('supply creates demand' and 'reduced supply reduces demand'). It is not possible to take this factor into account because:

- future development of supply is unknown (norms may not be realized); and
- **the degree of influence** which growth or limitation of supply exerts on demand cannot be determined;
- supply cannot be regulated in as far as autonomous developments are concerned (for instance initiatives deriving from the private sector).

The Background Study deals with the following facilities for the elderly:

- homes for the elderly and their neighbourhood function;
- nursing homes;
- day treatment in nursing homes;
- hospitals (general, special and university);
- district nursing;
- institutionalized home help;
- dwellings for the elderly; and
- co-ordinated work for the elderly.

The same facilities will be dealt with in the reference scenario, as well as care by GPs and ambulatory mental health care for the elderly. The order in which facilities are dealt with is more or less arbitrary. Both extra and intramural facilities are dealt with, and facilities fall both within the sphere of health care and of social services. Moreover, some facilities may be characterized as being typically facilities for the elderly, while others are more general facilities of which the elderly make use to a considerable degree.

Recently there has been much speculation in publications regarding the future use and costs of subsidized facilities. Various publications discuss several of the facilities mentioned above which are used by the elderly.[36)]
When these publications are compared, it is noticeable that the estimates differ considerably. A complex of factors accounts for this.

- The publications dating from before 1982 make use of the old CBS population forecast dating from 1976.
- Publications of after 1982 are based on different variants of the CBS population forecast of 1982.
- Use is made of various estimation methods;
  - null variant versus trend variant
  - flow model versus stock model (see further section 2.5.5).
- In the null variants estimates are based on different 'base years' (for instance sex and age-specific usage in 1978).
- The same 'calculation units' are not always employed: for instance number of beds in homes for the elderly versus number of patients.
- Some publications give only the 'financial results', without giving any clear justification for the estimation method used.
- When classification according to age is employed, often different age categories are used, or the age categories are not specified.

All these factors taken together have the effect of making estimates difficult to compare. In the text which follows, an overview per facility will be given of the available projections. Often these had to be supplemented with new calculations.

2.5.2 The development of facilities

At the end of the seventies there was a change in this pattern of growth. This change was not only related to the attempt to curb government expenditure, but also to changing goals in policy for the elderly.[37)]
At the moment the emphasis is on independence of the elderly and promoting their ability to manage for themselves, as well as postponement for as long as possible of admission to homes for the elderly or nursing homes. In order to attain this goal, every effort is made to shift the emphasis from intramural facilities (especially homes for the elderly) to extramural facilities (such as home help, district nursing, and co-ordinated work for the elderly). In the 'Note on Policy relating to the Elderly' written in June 1983 by the Minister of Welfare, Health and Culture and other Ministers, proposals are put forward for intensifying this policy.

Though there was an expansion of home help and district nursing in the second half of the seventies, there was no 'spontaneous' substitution. It is true that many of the elderly who were formerly dependent on aid from family and relatives or volunteer work, began to apply for extramural aid. Intensification of extramural care plus an underestimation of the supply of informal aid, have had the result that the demand for aid has shifted from informal aid to extramural aid.[38)]
The fact too that the share of costs of extramural care in the total between 1972 and 1982 decreased from 31 to 28% would seem to indicate that the envisaged shift from intra to extramural care has not been realized.[39)]

In order to be able to better specify the goals relating to policy for the elderly (and recently also as a consequence of the need felt for the limitation of the growth of government expenditure), capacity norms have been introduced with respect to most facilities. Sometimes these norms take the form of restrictions, while in other cases the accepted growth percentages for coming years are indicated. Policy with respect to these capacity norms has been rather subject to change in recent years, partly as a consequence of the changing economic situation and the event of a new cabinet in 1982.

In the following pages, the (norms) policy with respect to the facilities for the elderly selected by us are dealt with. This is confronted with future development of the demand for these facilities in the case of a constant level of care (null variant), and in the case of development according to observed trends.

2.5.3 GP care

By and large, the GP is the first professional provider of aid to be confronted with the health problems of the older patient. GP care occupies a central position in first echelon care. The GP is important as:
(a) provider of aid (in particular diagnosis and therapy);
(b) the person who refers patients to other facilities;
(c) contact person for other aid services.

Usually the number of contacts with the GP in a certain period is employed as indicator of the use made of GP care. The most important sources of information on this subject are at present the continuous CBS health survey, which is carried out since 1981, and the Continuous Morbidity Registration in Nijmegen.
The CBS health survey shows that the frequency of contacts since 1981 has declined slightly. This decline is found particularly among privately insured persons. However, no significant trend is observable.
Table 2.11 indicates the percentage of the population of 55 years of age and older who have been treated by the GP in a given year according to sex and type of insurance, as well as the number of visits per capita of the population and the number of visits per patient.

The percentage of very elderly people (75+) who have been treated by the GP is higher than for 'younger' elderly persons. When differentiated according to sex, however, there is definitely not a uniform increase with age. In every age category the percentage of women who visit the GP is greater than the number of men.
People insured under the health insurance fund are also found to visit the GP more often than privately insured persons, which might be an indication in differences in health between the two groups.[40)] This trend however does not persist in higher age groups (over 75 years of age).

Table 2.11 Number of contacts with the GP in one year, according to age category, sex and type of insurance: Health Surveys 1981 and 1982 (N=20,000)

age	total	men	women	Sick Fund.	Private Insur.
	Percentage of persons who contacted GP				
55-59	72.9	68.3	77.2	75.0	70.3
60-64	75.7	74.6	76.7	78.1	70.7
65-69	76.5	70.7	81.6	77.7	72.7
70-74	78.1	74.4	81.2	80.1	71.7
75 +	83.3	80.5	85.4	83.3	84.5
	Number of contacts per capita of population				
55-59	4.723	4.542	4.887	5.169	3.966
60-64	4.143	4.389	3.905	4.210	4.145
65-69	3.441	3.333	3.537	3.708	2.557
70-74	5.571	4.792	6.222	5.887	4.906
75 +	6.120	5.413	6.650	5.986	6.948
	Number of contacts per individual with contact				
55-59	6.474	6.653	6.331	6.896	5.639
60-64	5.474	5.885	5.088	5.390	5.867
65-69	4.500	4.718	4.333	4.771	3.519
70-74	7.133	6.440	7.663	7.351	6.842
75 +	7.344	6.720	7.784	7.188	8.224

Source: Monthly Bulletin Health Statistics (CBS) 84/10 p. 8 (adapted)

Since there has been no significant trend in the frequency of visits to GPs in recent years, for purposes of estimating future developments in the number of visits to the GP a constant age and sex-specific frequency may be assumed. The null variant then coincides with the trend variant. As a result of double ageing, in the future there will be an increase in the number of times the GP is consulted among the age group 55 years and older (see Table 2.12).

Table 2.12 Estimate of number of contacts with the GP (x 1.000) for the population of 55 years and older in one year (trend and null variants)

	absolute	number of contacts per capita of the population of 55 and older
1982	14,502	4.83
1985	15,122	4.86
1990	15,968	4.85
2000	17,915	4.89

Source: Trend and null variants:
* sex and age-specific calculations
* use is made of five age categories per sex, in accordance with the CBS classification (see Table 2.11.)

The increase in contact with the GP among older people is possibly to be explained by the fact that in general the complaints of older people are more severe than those of the younger elderly, and require more frequent medical attention, such as chronic-degenerative complaints (see also Appendix G).
Moreover, the **nature** of the contacts alters: more intensive contact and a shift from visits by the patient to the physician's surgery to visits by the physician to the patient at home.

In section 2.3 (The state of health of the elderly) it was stated that shifts in the pattern of morbidity result from (double) ageing. According to Van den Hoogen et al., in 2000 there will be an increase of 5% in the total number of complaints in the 'standard GP practice'. Table 2.4.a shows that the increase in complaints among the group over sixty years of age will be approximately 17%.
In how far GPs devote special attention to their older patients is not clear. No systematic studies on this subject are available. For preventative purposes, a few GPs carried out fairly extensive surveys among their elderly patients.[41)] The elderly patient sometimes poses specific problems in diagnosis and treatment.[42)]
Pleas are heard from various sources for more attention for geriatric problems in the training of GPs.
For instance:

- Adapted diagnostic criteria (normal values) to avoid **overdiagnosis** (for instance in the case of hypertension and diabetes mellitus).

- More attention for primary and secondary prevention (for instance early diagnosis in the case of dementia).

Other important functions performed by the GP are those of referring patients to other (health) care facilities and acting as contact person for other providers of aid, both for other facilities in first echelon care and with respect to second echelon care (outpatient care and treatment in hospital).
It is not possible to summarize briefly all the functions which the GP performs in relation to all the existing (health) care facilities. We will suffice with indicating a few more or less recent developments.

Government strives after reinforcing first echelon help, both **within the first echelon** (by means of intensification of cohesion and furthering co-operation) and **in relation to second echelon help** (by means of transferral of specific resources from intra to extramural care). The Memorandum on First Echelon Care (1983) contains the most recent policy proposals on this topic. For some time there has been a growth in co-operation between first echelon facilities (both health care facilities and social services).
Forms of co-operation in which GPs also participate are especially group practices and health centres. From 1970 to 1983 the number of group practices (joint practices of at least three GPs in one building) increased from 8 to 91. The number of health centres (a co-operative practice in one building consisting of at least one GP, one district nurse and one social worker) increased from 3 to 120. In 1983 the number of GPs in group practices was 302, and the number in health centres 375. This corresponds with 5.4% and 6.7% of the total number of independent GPs.[43)] Since 1970 the number of group practices and health centres has increased relatively quickly. The percentage of solo GPs thus also decreases proportionately. In 1970 85.4% of GPs had a solo practice. In 1983 this percentage had dropped to 57.9%. The percentage of solo GPs is lowest among the younger age categories and the age category 60-69 years of age.[44)]
Structured forms of co-operation are backed by government by means of subsidies.
According to the Memorandum on First Echelon Care, there is considerable discussion as to whether the provision of aid in (multidisciplinary) co-operation provides any advantages for patients. Surveys as to the relation between forms of co-operation and referral figures have shown that referrals in duo teams or group practices and health centres are somewhat lower than in solo practices (see Table 2.13).

Table 2.13 Number of referrals per 1000 Sick Fund patients according to type of practice

	Total	Solo practice	Duo practice	Group practice	Health Care centre
1978	458	464	441	449	436
1979	464	470	449	452	429
1980	466	473	451	460	423
1981	466	467	448	465	416
1982	477	486	462	462	447

Source: National Information System Sick Funds (LISZ), quoted in Memorandum on First Echelon Care, p. 27.

Also, the number of days spent in hospitals per 1000 health fund patients is found to be greater for patients from solo practices than for patients from joint practices (see Table 2.14).

Results of a study as to an explanation for these differences are not yet available.
One possible explanation might be that group practices are found mainly in 'younger' neighbourhoods; in other words the composition of the patients who make up the practice may exert a greater influence on referrals and number of days spent in hospital than the form of the practice as such. One advantage of co-operation might be that one of the workers (GP, physiotherapist) may be able to concentrate on the elderly.

There are still many obstacles to be overcome, such as attunement of tasks, expertise and working methods, when attempting to heighten co-operation in first echelon care. Meanwhile, the 'basic packet of tasks' has been drawn up and accepted by the National Society of General Practitioners.[45] Discussion, inter alia with financers, is still taking place on the matter.

As regard co-operation between GPs and second echelon health care, three experiments can be mentioned:
- the development of diagnostic facilities for GPs in two hospitals (Oudenrijn in Utrecht and St. Annadal in Maastricht);

- the possibility for GPs to obtain an opinion from a specialist as to diagnosis and treatment of a patient, who continues under the treatment of the GP;
- the possibility for GPs to have their geriatric patients examined once by colleagues trained in geriatrics. (This service is reserved for very old patients). For the time being, only the Nursing Home Heerlen offers this service.[46]

Table 2.14 Data covering number of referrals and number of days spent in hospital of registered Sick Fund patients in 1978

	solo GP	solo + assist.	associates	group practices	health centres
Number of referrals per 1000 Sick Fund patients	467	462	457	454	425
Average number of days spent in hospital per admission	16.7	16.5	16.6	16.4	16.1
Number of days in hospital per 1000 Sick Fund patients	1938	1907	1908	1847	1776

Source: Survey as to Co-operation and Referrals, Part 1: Nature of Practice and Production Figures; Dutch Institute of General Practitioners, 1983. Quoted in Memorandum on First Echelon Care, p. 26.

The above-mentioned experiments call for structured co-operation, clearly defined agreements, and consultation procedures between GPs and second echelon facilities.
In the future, further expansion is to be expected in co-operation within the first echelon sector and between this sector and other sectors of care, though this will mean that it will be necessary for the various sectors to relinquish some degree of the autonomy to which they have been accustomed. GP practices may in this context be characterized as key facilities which will be further reinforced.

2.5.4 Ambulatory mental health care

Mental health care of the elderly encompasses a wide range of facilities. This section deals with extramural mental health care, which consists of the assistance provided to the elderly by the Regional Institutes for Ambulatory Mental Health Care (RIAGGs).

Within mental health care, ambulatory mental health care (AGGZ) constitutes a system of facilities that on the one hand encompasses a number of specialistic treatment methods of its own (and which as such fall under second echelon facilities), and on the other hand performs an intermediary function between non-specialized first echelon health care and specialized second echelon facilities. Though AGGZ is fairly accessible, most patients nonetheless come into contact with AGGZ via referrals from GPs. This is especially true with respect to the elderly.
The social-psychogeriatric function of the AGGZ is aimed at the older patient, who as a result of disturbances of the physical/psycho-social balance, to a greater or lesser degree evinces behavioural disturbances.[47] The central goal is to make it possible for the elderly to remain in their own surroundings for as long as possible. The AGGZ aims at furthering the efficient functioning and co-ordination of aid systems, so that admission to intramural facilities can be avoided or at least postponed.
The main functions of the AGGZ in the area of care of the elderly are:

1. prevention;
2. providing aid: placing its own packet of services, inclusive of diagnosis, at the disposal of the patient; acting as intermediary in the case of admission and discharge from intramural and semimural facilities (nursing homes, day treatment, homes for the elderly, psychiatric or other wards of general hospitals, etc.);
3. provision of services: consultations and advice, especially to first echelon health care (inclusive of GPs).

The organization of social-geriatric functions within the AGGZ is in the process of development. The goal is to create separate departments of teams within the social-geriatric services of the RIAGGs. At the moment, 54 of the 59 recognized RIAGGs have a separate organizational set-up for psychogeriatrics. In a few large cities, psychogeriatrics fall under the Municipal Medical and Health Service. The goal is to ensure that every RIAGG has at least one such team at its disposal. This goal is consistent with the policy for reinforcement of first echelon (health) care. Together with for instance nursing homes, the social psychogeriatric service may fulfill a central role, provided it has at its disposal a complete multidisciplinary team, consisting of a geriatrician, a psychogeriatric nurse, and at least one psychologist and social worker.

In view of the inadequate data with respect to developments in the use of the AGGZ in the past, no trend variant for future use can be calculated. On the basis of usage figures for 1982 it is however possible to estimate future demand according to the null variant. Table 2.15 gives the age and sex-specific usage of the AGGZ in 1982.

Table 2.15 Number of patients of 65 years of age and older according to age and sex (absolute and % of the population of 65 years of age and older) registered with Ambulatory Mental Health Care in 1982[48]

age	male		female	
	abs.	%	abs.	%
65-69	1301	0.53	2007	0.67
70-74	1408	0.75	2516	0.96
75-79	1741	1.35	3409	1.66
80-84	1619	2.26	3365	2.58
85 +	1856	4.05	3939	4.47

Source: Data of the Dutch Assn. for Ambulatory Mental Health Care

In coming decades, the double ageing will mean that the AGGZ will be called upon to provide an increased amount of psychogeriatric care. Table 2.16 gives the estimate of this demand according to the null variant.

This projection can only be regarded as a rough estimate. One of the reasons for this is that in addition to age and sex, 'civil status' constitutes an important distinguishing characteristic: the percentage of divorcees who appeal to the AGGZ is about treble that of the remainder of the population.[49] Appendix B shows that, according to the CBS forecast, the number of divorced persons over the age of 65 will increase from 43,000 in 1980 to 119,000 in 2000. In view of the general trend, older divorcees too will make greater demands on the AGGZ.

Table 2.16 Estimate of the number of patients of 65 years and older registered with Ambulatory Mental Health Care in the period 1982-2000, absolute and in % of the population of 65 and older (null variant)

	abs.	%
1982	23,161	1.39
1985	24,811	1.43
1990	27,631	1.46
2000	41,043	1.96

Source: Age and sex-specific calculations on the basis of usage percentages from Table 2.15.

As regards the fixing of norms with respect to the supply of psychogeriatric services within the AGGZ (the 'policy variant'), the Health Fund Council considers 15 new requests for assistance per 1000 elderly persons per annum to be the maximum. Since 1.1.1982 the AGGZ has been financed under the Exceptional Medical Expenses Act. As a consequence of this, the RIAGGs themselves are in a position to decide how the budget at their disposal will be divided over the various categories of patients. As guideline, the 20-20 norm is adhered to. That is to say, calculations are based on requests for assistance from 20 new elderly persons per 1000 elderly persons in the population per annum, and on 20 hours of assistance for each person.
At present, discussions are in progress as to differentiation of this norm. In practice it has been found that the very old (who increasingly appeal to the AGGZ for aid) require more intensive help: approximately 40 hours per patient. It is however to be questioned whether the demand for aid really corresponds with the actual need for assistance.[50)]

Till the year 2000, a further intensification of the social-geriatric function of the RIAGGs is to be expected. Most likely each RIAGG will have at its disposal a separate department or team in the form of a social geriatric service (SGD). The position of these SGDs within the first echelon will be enhanced, in spite of the problems attaching to increased cohesion and co-operation between facilities.[51]

In the framework of regionalization policy, the envisaged expansion of the role of the AGGZ (shifts in capacity and functions), will come about via the creation of a network of Regional Institutes for Mental Health Care (RIGGs) into which the RIAGGs will be incorporated. Over a phased period, at least one RIGG will be created in each health care region.[52]

2.5.5 Homes for the elderly

The most recent figures regarding the number of residents in homes for the elderly date from 31.12.1980.[53] On that date, the number of residents was 135,242. Of this number 1919 were younger than 65 (1.4%). This means that there were 133,323 persons of 65 years of age and older in homes for the elderly. This is 8.1% of all persons of 65 years of age and older in the Netherlands.
The percentages of people in these homes climb rapidly per age category. See Table 2.17. At the end of 1980, of the total number of residents of homes, 34.4% accrued to the age category 65-79, and 64.2% to the age category 80 years and older.

In recent years, the average age of residents has climbed sharply to over 80, partly because the age at which a person enters such a home has raisen.
Since 1977 the 7% norm applies for homes for the elderly; this means that the total capacity of homes for the elderly must be such that 7% of the people of 65 years and older can be placed in a home. Neither the correctness nor the incorrectness of this norm has ever been demonstrated.[54]
On the basis of demographic developments, in the future an increase of this norm would be necessary. This will hold good both in the event that the level of care remains constant and in the event that account is taken of development trends in the past. The calculated percentages range from 9.39% (Note on Policy Relating to the Elderly - constant level of care on the basis of 1978) to 7.62% (Pommer and Wiebrens - trend variant[55]) in 1990.
At the moment, as we said above, the average age of residents of homes for the elderly is slightly over 80. Recent survey data show that the average age of persons who have been declared eligible for admission is somewhat higher.[56]

Table 2.17 Population of 65 and older and elderly persons of 65 and older resident in homes for the elderly, according to age and sex, December 31st, 1980[a)]

men	total[b)]	resident in homes for elderly absolute		%
65-69	243,238	1,415	(0.6)	0.6
70-74	187,683	3,994	(2.2)	2.1
75-79	127,511	8,334	(6.8)	6.5
80-84	70,546	11,486	(16.6)	16.3
85-89	32,224	9,078	(28.5)	28.2
90 +	12,420	4,451	(35.5)	35.8
total	673,622	38,758	(5.9)	5.8
women				
65-69	298,867	2,820	(1.0)	0.9
70-74	260,578	9,078	(3.7)	3.5
75-79	201,039	20,882	(10.9)	10.4
80-84	125,223	29,284	(24.1)	23.4
85-89	60,286	22,615	(37.6)	37.5
90 +	22,392	9,886	(43.4)	44.2
total	968,385	94,565	(9.8)	9.8
men and women				
65-69	542,105	4,235	(0.8)	0.8
70-74	448,261	13,072	(3.1)	2.9
75-79	328,550	29,216	(9.3)	8.9
80-84	195,769	40,770	(21.4)	20.8
85-89	92,510	31,693	(34.4)	34.3
85-89	34,812	14,337	(40.5)	41.2
total	1,642,007	133,323	(8.2)	8.1

(a) In brackets the figures for 1979
(b) Population per December 31st, 1980

Source: CBS, 1983, p. 13.

The heightening of the average admission age has also resulted in an increase in the amount of care required. This involves consequences for the quality of care and nursing in homes for the elderly. The Note on Policy Relating to the Elderly takes special account of the possibility for future co-operation between homes for the elderly and nursing homes, as a result of which patients who, viewed objectively (on account of their need for nursing) would otherwise have to be transferred to a nursing home, would be able to remain in a home for the elderly. Many uncertainties however still attach to organizational, staff and financial aspects.

The most recent policy pronouncements regarding the future capacity of homes for the elderly are laid down in the Note on Policy Relating to the Elderly of June 1983, and Report 45 of the Committee for Reconsideration of Collective Expenditure (Reconsideration of Policy relating to the Elderly, Second Chamber, 1981-1982 Session).

On the one hand, the Note poses that 'the range of relevant intramural facilities will be adapted as adequately as possible to existing norms', but on the other hand it also states that 'in the future there will be a minimum of expansion of the number of places in homes for the elderly'. Finally the Note refers to 'a constant capacity of beds in homes for the elderly'. We consequently assume that there will be a halt to the construction of homes for the elderly. In order to stem the flow of the elderly to homes in the future, the Note proposes that the system of indication for admission be revised, and that certain urgency categories be dropped. It would however appear that if certain categories of urgency are dropped, the elderly who now fall within these categories will in general be placed under other urgency categories. This is caused by an inadequate indication system, and the great need for assistance among the elderly (over 80 years of age) who apply for admission to homes.[57)]

Table 2.18 gives results of estimates as to the number of places which will be required in homes for the elderly in the future. The first column (null variant, that is to say constant level of care) has been calculated on the basis of the most recent sex and age-specific usage percentages. In the trend variant, use was made of a so-called 'flow model' instead of a 'stock model'.[58)]

Table 2.18 Estimate of the capacity of homes for the elderly (absolute and in percentages of the population of 65 and older) in the period 1981-2000

	null variant		trend variant		policy variant	
	abs.	%59)	abs.	%	abs.	%
1.1.1981	133,000	8.1	133,000	8.1	145,000	8.8
1.1.1985	151,000	8.7	136,000	7.9	145,000	8.4
1.1.1990	170,000	9.0	145,000	7.7	145,000	7.7
1.1.2000	196,000	9.4	60)		145,000	6.9

Source: **Null variant:**
Application of age and sex-specific usage figures from Table 2.17 to the population of 65 and older in 1985, 1990 and 2000 resp. The figure refers to the number of patients to the nearest thousand.
Trend variant:
Pommer and Wiebrens, 1984, pp. 88-89. Here too figures refer to patients to the nearest thousand.
Policy variant:
Constant number of beds. Very recently however, (September 1984), the Minister approved 800 (new) beds for nursing homes in the year 1984. Figures given in the Table refer to the number of beds, to the nearest thousand.

If the present building stop were to persist till the year 2000, the present norm of 7% would be fulfilled. Figure II.5 shows the results for the null and the policy variants in graph form.

One of the instruments for furthering the independance and social integration of the elderly is the district function of homes for the elderly. The following eleven district-oriented functions of homes are quoted in the Note on Policy Relating to the Elderly:61)

- provision of meals;
- monitoring by means of alarm systems;
- first aid function in case of emergency;
- use of bath;
- assistance with baths and hair-dressing;
- chiropodist services;
- socio-cultural activities;
- keep-fit classes for the elderly;

- day-care in cases of immediate need;
- night-care in cases of immediate need;
- short-term care in the case of illness.

Figure II.5 Number of beds and patients in homes for the elderly, 1984-2000, null and policy variants.

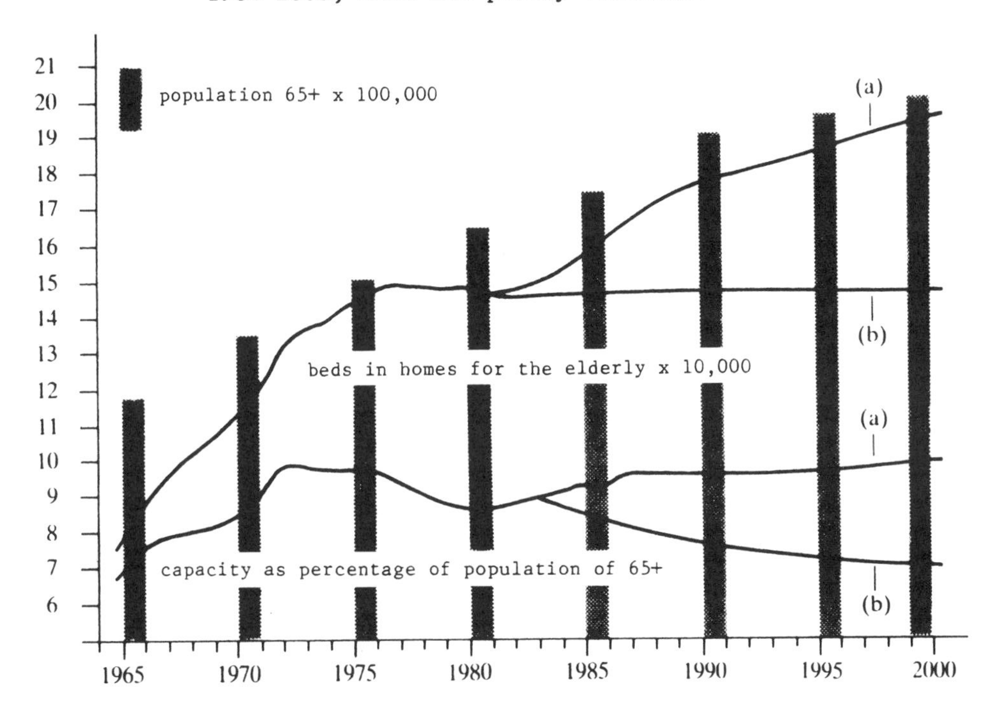

(a) Null variant
(b) Policy variant

A lot of uncertainties still exist with respect to the development of the district function of homes for the elderly. Some homes have already for years served a neighbourhood function, for instance some new homes in new towns. Other homes have only recently begun to orientate more to the neighbourhood. What will be the ultimate neighbourhood function of homes is by no means definitely outlined, and is to a large extent dependent on local circumstances, for instance the degree to which residents of homes are representative of the residents of the neighbourhood, the situation of the home, infrastructure, spatial possibilities in the homes, etc.

Concluding, it may be said that the reference scenario proceeds from the assumption that the residential function of homes for the elderly will decline till the year 2000, and that the emphasis will shift much more to the nursing function and the district-orientated function. This is in accordance with the goals with respect to policy for the elderly presently pursued.

2.5.6 Nursing homes

Nursing homes include homes for somatically ill patients, homes for psychogeriatric patients and combined nursing homes. As of 1.1.1984, there were approximately 47,500 nursing home beds, divided over 148 homes for somatic patients, 83 for psychogeriatric patients and 96 combined nursing homes.
In the null variant, the future use of nursing homes has been estimated by applying the most recent sex and age-specific usage figures (1983) to the population of 65 years of age and older in the years 1985, 1990 and 2000 respectively. For the trend variant, the estimate is based on development trends on sex and age-specific use in the past. Among the elderly usage shows a levelling off of growth.[62)]

In the policy variant use is made of present norms.
According to Paragraph 3 of the Hospitals Facilities Act, the following number of beds are prescribed in each type of nursing home:

- for somatic patients: 1.2 beds per 100 inhabitants in the age category 65 years of age and older, plus 0.35 beds per 1000 inhabitants of the total population;
- for psychogeriatric patients: 1.25 beds per 100 inhabitants in the age category 65 years of age and older (for the municipalities of Amsterdam, Rotterdam and The Hague raised to a maximum of 1.5% of this age category).[63)]

In 1980 the number of recognized nursing home beds for patients with somatic diseases was approximately 3000 more than according to the prescribed norm, and had already reached the norm for 1990. Consequently there is a building stop in this sector. As against this, the number of recognized psychogeriatric nursing home beds was 2400 lower than the capacity prescribed for 1980.
Bed facilities for 5000 psychogeriatric patients are either in the process of construction or preparations are being made for starting construction. When these facilities have been completed, it will on

the basis of existing guidelines be considered that there are sufficient beds for the two categories of patients till 1990. Table 2.19 present estimate results for the three variants.

Table 2.19 Estimate of the capacity of nursing homes (somatic, psychogeriatric and combined) 1983-2000, absolute and % of 65-plussers

	null variant		trend variant		policy variant
	abs.	%	abs.	%	abs.
1.1.1983	47,456	2.80	47,456	2.80	46,471
1.1.1985	49,410	2.86	51,743	2.99	47,457
1.1.1990	55,169	2.91	59,053	3.12	51,669
1.1.2000	61,248	2.93	60)		56,656

Source: Null variant:
* Data per 1.1.1983: Hospitals Institute of the Netherlands: institutions for intramural health care. Basic data 1.1.1983.
* Use is made of three age categories per sex, in accordance with the classification adhered to by the SCP.
* Usage percentages men:
  0.036 for 0-64-year-olds
  0.98 for 65-79-year-olds
  4.93 for 80-year-olds and older.
* Usage percentages for women:
  0.043 for 0-64-year-olds
  1.47 for 65-79-year-olds
  8.49 for 80-year-olds and older.

Trend variant:
Pommer and Wiebrens: Costs and Financing of Facilities for the Elderly, 1981-1991. SCP Working Paper No. 39, 1984, pp. 86-87.

Policy variant:
In accordance with the norm laid down in the Hospital Facilities Act (also for 1983).

The norm of 1.25% for psychogeriatric patients is however considered too low. The age-specific usage figures within the group of elderly in combination with demographic developments make it essential that

the norm be raised to at least 1.5%. According to the Note on Policy Relating to the Elderly, without this increase it will not be possible to maintain care of nursing home patients at the present level.[64]
The vast majority of nursing home patients consists of women (more than two-thirds of somatic patients and nearly three-quarters of psychogeriatric patients). This overrepresentation of women can partly be attributed to demographic causes: there are more (very) old women than (very) old men, and of the very old, more women than men live alone. In general, possibilities for nursing (very) old people who reside alone at home are limited.

In spite of the fact that the average duration of life of patients in nursing homes is less than two years, lengthy sojourns especially by psychogeriatric patients are by no means unusual. On the average these patients are also older than the somatic patients.[65]

In recent years, the role of the nursing home as such has come under discussion. Apart from the costs aspect, the motives for this discussion are in particular altered values and norms with respect to privacy and individualization on the one hand, and the pursuit of social integration (averting isolation) on the other hand.
To keep pace with these developments, nursing homes have begun to introduce greater differentiation in the services offered, for instance stressing reactivation and treatment by comparison with nursing as such.[66] At present nursing homes are opening up various possibilities, such as admission for the night, admission for the week-end, day-treatment, and intervention in crisis. Data for the country as a whole are scarce regarding these more recent functions. For the present it is difficult to predict how these new forms of admission will develop qualitatively and quantitatively.
It may be assumed that in the future the differences between nursing homes and homes for the elderly will become less, partly as a consequence of the more rigid requirements for admission to homes for the elderly. Initiatives are already being taken for co-operation. This process can be speeded up inter alia by placing financing of both types of institution on a more equal footing.

2.5.7 Day treatment in nursing homes

By day treatment is meant making use of the possibilities offered by the nursing home, without being permanently admitted. The patients concerned are eligible for admission, but evening, night and week-end care takes place wherever the patient resides. Day treatment aims at reactivation and resocialization in order to avoid or at least to postpone permanent admission to the nursing home.

As of 1.1.1984, there were 686 recognized day-treatment places in somatic nursing homes, 510 in psychogeriatric and 1029 in combined nursing homes (a total of 2225).[67] The Memorandum of Health Policy With Limited Resources indicates that capacity will be expanded if this appears desirable on the basis of the results of current evaluation studies.[68] Past studies as to day-treatment have however shown that there are no clear substitution effects.[69]
On the one hand, users of day-treatment services are already in homes, while on the other hand in practice day-treatment often serves the purpose of relieving the situation at home. In this manner, new categories of patients are reached.
The norm for the number of day-treatment places is applied to the forecasted population per health region. For somatic patients this norm is 0.72 places per 1000 inhabitants in the age category 65 years of age and older plus 0.02 places per 1000 of the total population. For psychogeriatric patients the norm is 0.75 places per 1000 inhabitants in the age group 65 years of age and older.[70]

If no account is taken of the envisaged regionalization, and the norms quoted are applied to the population of the country as a whole, the following picture of the policy variant is obtained (since this facility has been in existence only for a relatively short time and is still very much in the stage of development, it would not serve much purpose to make an estimate of the numbers of places according to the null and the trend variants).

Table 2.20 Estimate of the number of day-treatment places (absolute) in the period 1984-2000 (policy variant)

1984	2803
1985	2833
1990	3084
2000	3384

Since the actual number of day treatment places on 1.1.1984 was only 2225, an expansion of approximately 580 places would be possible before the norm is achieved. At the moment it is not clear whether this expansion will actually take place should the results of the above-mentioned evaluation study be negative for the facility 'day treatment'.

2.5.8 Hospitals

This facility encompasses general, university and special hospitals. This is a general facility used by all age categories. Treatment offered can be either intramural or extramural (outpatient departments).
When we take a look at intramural treatment, it is found that the three discharge diagnoses most often encountered in patients of 55 years of age and older are tumors, coronary and arterial diseases, and diseases of the alimentary system (see Appendix H-4; cf. also Appendices H-1 to H-3).

Development trends in demand per age category are varied. According to calculations of the Social and Cultural Planning Bureau (SCP), in the period 1981-1991 there will be a decrease of 6.9% in the number of days spent in hospital. The decrease is very marked among the age categories 0-44 years of age, and moderate among the 45-64 year olds. Among the 65-plussers, the trend is towards an absolute increase in the number of days spent in hospital.[71] The relative trend shows a slight decrease (see Table 2.21).
In hospitals too, government is making efforts to reduce the number of beds. The Guidelines in Paragraph 3 of the Hospitals Facilities Act propose a norm of 3.7 beds per 1,000 inhabitants.[72] This norm should be realized in 1988. As of 1.1.1982 there were 61,600 recognized beds to which the norm applied, which is equivalent to 4.3 beds per 1000 inhabitants.
According to the SCP, the effect of the 3.7 per thousand norm on the number of elderly persons in hospitals is difficult to assess. In the first place, it is not known to what degree the desired reduction of the number of beds will affect the elderly and other citizens. In the second place, the question remains as to whether the reduction of the number of beds will result in a reduction of number of days spent in hospital or of the number of patients. If it is assumed that the reduction of beds will affect the elderly and the remainder of inhabitants proportionate to present usage, this would lead to the estimates in the policy variant given in Table 2.21.

Table 2.21 Estimate of number of days spent in hospital by people in the age category 65 and older in the period 1982-2000 (total and per person)

	null variant		trend variant		policy variant
	abs.	d/p	abs.	d/p	abs.
1982	7,772,000	4.63	7,722,000	4.63	7,722,000
1985	7,995,000	4.62	7,827,000	4.52	-
1990	8,761,000	4.62	8.520.000	4.50	6,308,000
2000	9,664,000	4.63	60)		6,429,000

d/p = number of days in hospital per person

Source: Null variant
* sex-specific, but not age-specific calculation
* usage figure 1982: M=3,364,000 days and F=4,358,000 days (65 and older)

Trend variant
* Pommer and Wiebrens, 1984, pp. 90-91.

Policy variant
* number of days in hospital for population of 65+: 35.8% of the total number of days in hospital
* level of occupancy: 85%
* number of days in hospital 65+ in 1990: 3.7 x 15,350 x 0.85 x 365 x 0.358
* number of days in hospital 65+ in 2000: 3.7 x 15,643 x 0.85 x 365 x 0.358
* number of days in hospital 65+ in 1985 cannot be calculated since the norm will only be reached in 1988.

In a section on the facility 'hospitals' we should not pass over the question of the development of new geriatric departments in general hospitals (GAAZs).
At the moment there are five GAAZs. These are all in hospitals in regions with a considerable overcapacity of beds.[73] It is uncertain how this number will develop till the year 2000. For the time being, government maintains a conservative policy.
Meanwhile, six experiments have been started. It is investigated whether it would be possible to introduce differentiation in the points of departure for these experiments. These experiments are being evaluated by the Hospitals Institute of the Netherlands in co-operation with the University of Groningen. The target group at

which the GAAZs are directed are (very) old people who are characterized by multiple pathology, a threatened permanent need for assistance, and disturbances in at least two of the three areas of functioning: physical, mental or social. The GAAZs aim at adequate diagnosis (i.a. by means of screening) and adequate referral. Outpatient treatment in hospitals has recently acquired extra significance for the elderly as a consequence of the appearance of the geriatric outpatient department (for instance in the Slotervaart Hospital). There are still only a limited number of these departments. It is uncertain how the number of geriatric outpatient departments will develop in the future.

2.5.9 District nursing

Data on district nursing are scarce. Adequate usage indicators are lacking, while it is moreover difficult to distinguish district nursing of the elderly from other duties carried out by Cross Organizations (e.g. infant care, general prevention). Data as to the development of the number of hours of district nursing are scarce, but would seem to indicate that the per capita use made of district nursing is stable.[74] As a result of this, the trend and null variants coincide. Figures concerning the intensity of age-specific usage have been taken from the Supplementary Survey on the Use of Facilities 1979 of the SCP. For the population of 65 years of age and older, in 1979 the per capita number of hours of district nursing was 4.085. For the age group 50 to 64, the number of hours was 0.285. In our estimates we limit ourselves to the age group 65 years of age and older.[75] See Table 2.22.

Recently, the National Association of Cross Organizations draws up estimates covering several years. The most recent figures concern the period 1985-1988. In these estimates, the consequences of double ageing for the number of people receiving care are indicated.[76]

The percentages of elderly people who now receive care from Cross Organizations are taken from registers in a number of 'basis units' which make use of a new district administration system drawn up according to age category.[77] The percentages of persons receiving care are:

- 60-69 year olds 9%
- 70-79 year olds 23%
- 80 years and older 40%

Table 2.22 Estimate of number of hours of district nursing for population of 65 and older (trend and null variants)

	trend and null variant		policy variant
	absolute	hrs/person	
1979	6,529,000	4.085	In the period 1984-1986:
1985	7,063,000	4.085	4% volume growth. Norm:
1990	7,741,000	4.085	1 district nurse per
2000	8,534,000	4.085	2500 inhabitants

Source: Trend and null variants:
Ministry for the Interior/SCP, 1983, p. 165 (for 1979 only).
Policy variant:
Memorandum on Health Policy With Limited Resources, 1983, p. 24 and Memorandum on First Echelon Care, p. 50.

The increase of the number of persons receiving care is estimated at 25,500 for the period 1985-1988.[78] This estimate proceeds from an average of 20 contacts per person per annum with an average duration of 40 minutes per contact. This amounts to slightly more than 13 hours per person per annum. It is not possible to make an adequate comparison with Table 2.22 since the classification into age categories does not coincide. If however, we assume that usage in the age group 65-69 years of age is also 9%, we arrive at a total of 5,313,000 hours of aid in 1988.[79] From this it is to be seen that such calculations are very sensitive to registration methods.

Limitation of the capacity of intramural facilities will have consequences for, inter alia, district nursing. The SCP has designed a substitution model in which the effects for district nursing and home help of the fixing of norms for hospitals, nursing homes and homes for the elderly are assessed.[80] Figure II.6 outlines this substitution model.

The coefficients in the model show the fraction of the number of elderly persons who will have to fall back on other facilities if the flow to intramural institutions is restricted. The coefficients are partly based on research, and partly on assumptions.[81]

Figure II.6 The substitution model

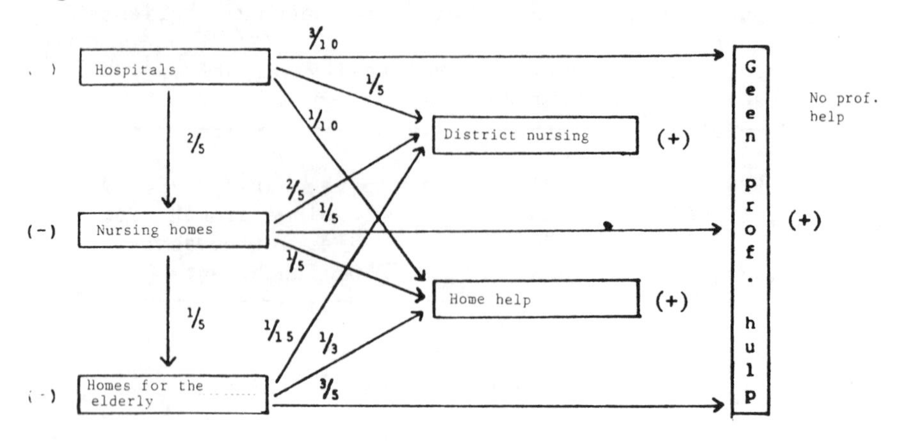

Source: Pommer and Wiebrens, 1984, p. 47.

Figure II.6 shows that 1/5 of the patients (65-plussers) who become the 'victims' of the reduction of beds in hospitals, will be obliged to appeal to district nursing. In the case of nursing homes this proportion is 2/5, and in the case of homes for the elderly 1/15.

The estimates of the National Association of Cross Organizations are vague with respect to these substitution effects. Account is taken of a possible 'increase of work load' as a consequence of differentiation of forms of care in nursing homes. Due to this differentiation, there will be an increase in the share of patients 'on the district' who need intensive nursing.[82)]
The altered admission and discharge policy of hospitals will have the same effect. According to the National Association of Cross Organizations this often necessitates complicated medical and nursing aid. Another recent development is the pursuit of 24-hour availability. Apart from the above, a number of social developments can be mentioned which result in the need for intensification of (district nursing) care:[83)]

- socio-economic developments such as unemployment, increasing loneliness and divorce make people more vulnerable;
- the socio-cultural diversity intensifies the demand for differentiation in approach: among the ethnic minorities for instance there is already an ageing first generation;
- more attention to preventative care for the elderly (i.a. group aid and counselling).

Government too envisages a growth of the work of the Cross Organizations in connection with the envisaged shift from intra to extramural aid. In 1982 a growth in volume of 6% was achieved, but as a result of financial-economic problems this was reduced to 4% for 1983. In the immediate future, Cross Organizations will be exempted from economic cuts.[84]
At the moment it is not at all clear how the work of the Cross Organizations will develop long-term. The norm which is strived after is one district nurse per 2500 inhabitants.[85]

2.5.10 Home help

Home help increased considerably in the seventies. In accordance with the Memorandum on Policy Relating to the Elderly 1970, this increase has been especially noticeable among the elderly. In the period 1970-1975 the average annual growth was slightly more than 15%. In the period 1976-1981 the annual growth was approximately 5%.[86] More than half of home help is to the elderly.
Estimates of future use of home help are almost identical for the null and trend variants (both absolutely and relatively). See Table 2.23. The picture which evolves is one of decreasing growth.

Table 2.23 Estimate of use of home help in total number of hours of aid for the elderly (x 1000) and in number of hours of aid per elderly person in the period 1980-2000 (null variant and trend variant)

	null variant		trend variant	
	abs.	hrs/person	abs.	hrs/person
1980	28,553	17.70	28,553	17.70
1985	31,473	18.20	31,481	18.21
1990	34,221	18.06	34,137	18.01
2000	38,419	18.39	60)	

Source: Null variant:
* Age specific but not sex-specific calculation
Trend variant:
* Source: Pommer and Wiebrens, 1984, pp. 84-85.

Home help too will feel the consequences of a limitation of the flow to intramural facilities. The effects of this can be seen in Figure II.6.

As yet, no policy norm has been laid down with respect to home help. In 1983 the average amount of home help was 25.7 hours per elderly person living at home. According to the Memorandum on First Echelon Care, further study and consultation will be necessary to evaluate needs. The level for 1983 is considered a guideline. Incidentally, a calculation according to the norm of 25.7 hours yields a (considerable) discrepancy with the estimates in the null and the trend variants. The point of departure when drawing up these estimates was the level of use in 1980, when per capita assistance of the population of 65 years of age and older was 17.7 hours (compare Table 2.23 with Table 2.24).

Table 2.24 Estimate of the use of home help in number of hours of aid (x 1000) according to norm of 25.7 hours per independently residing elderly person (policy variant)

Year	Hours of aid (x 1000)
1985	39,592
1990	43,393
2000	47,835

Source: * Percentage of independently residing elderly persons: 89.1% in 1983 (8.1% reside in homes for the elderly and 2.8% in nursing homes)

* The percentage of independently residing elderly persons is maintained constant till 2000. In reality this percentage will rise in connection with the limitation of influx to homes for the elderly to approximately 7% of the population of 65+. As against this, the percentage admitted to nursing homes will rise somewhat in connection with the expansion of the number of places for psycho-geriatric patients.

If it is desired to maintain the quality of care, either the same quality of care will have to be provided in less hours, or the costs per hour of assistance must drop. An attempt is being made to achieve the latter by employing 'alpha helpers' who are considerably cheaper (these provide domestic help only). In addition to this, according to the Note on Policy Relating to the Elderly, solutions must be sought in compensating measures:

(1) More attention for engaging volunteers and for self-care and umbrella care.
(2) Co-operation with other extramural facilities, especially district nursing, in order to avoid the overlapping of tasks, with, in addition to this, co-operation with second echelon facilities, including homes for the elderly and nursing homes.
(3) Temporary intramural care in cases where lengthy and intensive care is required.
(4) Better organizational attunement to the altered demand for assistance, with, inter alia, distribution of the number of hours of aid over the week.
(5) More attention for new forms of assistance, such as district care of the elderly.

The Memorandum Flanking Policy for the Elderly (1983) goes in detail into the adaptation of home help to the increasing requests for assistance from people who are heavily dependent on aid. The present budget offers scope for adapting the supply of aid to demand in as far as this concerns permanent availability, a permanent supply of aid, and availability in emergency. The Memorandum however states that should there be a disproportionate increase in the number of people heavily dependent on assistance, a relative expansion will be required.

2.5.11 Dwellings for the elderly

One of the instruments for realizing the goals of policy with respect to the elderly is to provide adapted forms of housing. The type of housing envisaged is something between the normal dwelling house and an institution, adapted to the needs of older residents as regards design, technical facilities and often also provision of services. These dwelling forms show very large individual differences as regards level of facilities.

The following types of dwellings can be distinguished:[87)]

- dwellings, which as regards design, situation and technical facilities are considered suitable for healthy elderly persons. These are ordinary small dwellings with three rooms or less;
- dwellings for the elderly which are adapted to the impaired functioning of the inhabitants. They should be situated at a short distance from public facilities and aid services, while

they should also contain an alarm system;

- monitored dwellings: these meet the same criteria as the above, but are built on the premises of or near to an institution (a home for the elderly). Services such as meals and household assistance can be provided from the institution;
- service flats: purchase service flats (intented for the more financially independent) and rented service flats often bear some similarity with an institutional home for the elderly. Services offered vary to some extent, but collective facilities are usually present, and domestic help is provided.[88]

The three latter types of dwellings constitute only a relatively small share of the dwellings of independently residing elderly persons. By far the majority of elderly people live in ordinary dwellings. Table 2.25 gives an overview of the residential situation.

This Table does not show what percentage of the elderly live in dwellings suited to their needs. The Housing Demand Survey of 1981 provides more information. Of all households the head of which is 55 years of age and older, 22% live in suitable dwellings which accrue to one of the four above-mentioned types.
Officially, the number of dwellings for the elderly in the Netherlands is 110,000. Of people who live in residences particularly adapted to the needs of the elderly, 15% live in a monitored dwelling (a complex of residences near to an instituationalized home). The type of dwelling for elderly citizens most often encountered is a complex of 10 or more dwellings without facilities (a third of all adapted residences for the elderly).

For the Netherlands as a whole, at the end of 1981 there was a shortage of nearly 110,000 suitable dwellings for elderly citizens. Half of this shortage is to be found in the Western part of the country.
In the recent Housing Demand Survey of 1981 as to residential requirements, a forecast was made of the need for dwellings suited to the elderly in 1990 and 2000. The estimate proceeded from a constant demand per age category,[90] based on percentages found in 1981.
In addition to the existing shortage, this will constitute a further demand of 10,000 suitable dwellings per annum between 1980 and 1990 and a demand of nearly 7,500 per annum between 1990 and 2000.[91]

Table 2.25 People living independently according to type of dwellings and age category in percentages in 1982

age category	ordinary dwelling	adapted dwelling	service flat	other[89]
men				
55-59	98	1	0	1
60-64	98	1	0	1
65-69	95	3	1	1
70-74	91	6	2	0
75-79	87	10	3	-
80 a.o.	76	16	6	2
women				
55-59	99	1	0	1
60-64	96	3	0	1
65-69	93	6	2	0
70-74	82	11	4	1
75-79	78	15	6	1
80 a.o.	69	22	8	0
men and women				
total				
55 a.o.	91	6	2	1
65 a.o.	86	10	4	1

Source: CBS, 1984, p. 48 (adapted)

The Housing Demand Survey of 1981 shows that in that year 90% of households consisting of persons of 55 years of age and older reside in an independent residence (for the age group 65+ this percentage was 85).[92]
In recent years, this 'degree of independence' has increased, and will increase even further in the future as a result of policy, especially with respect to the limitation of the number of places in care institutions. For every three places in care institutions which do not materialize, it will be necessary to provide approximately two extra dwellings suited specially to the needs of the elderly. This means that approximately 2000 extra dwellings per annum will be required.[93] Table 2.26 gives an overview of the above.

Table 2.26 Estimate of necessary increase in number of suitable dwellings for the elderly 1981-2000 (null variant and substitution effect)

	null-variant		substitution effect	total
1981	110,000	(shortage)	--	110,000
1981-1990	100,000		20,000	120,000
1991-2000	75,000		20,000	95,000

At present there is no 'policy norm' with respect to suitable housing for independently residing elderly people.
According to the Note on Policy Relating to the Elderly, in addition to the construction of new dwellings and adaptation of existing ones for the elderly, a specific policy relating to distribution of dwellings would constitute an effective instrument for providing adequate housing for the elderly. A comparison of the housing demand survey of 1981 with that of 1977 however shows that a fairly large number of dwellings which would be suitable for the elderly are occupied by people younger than 55 years of age[94)] and that in the period 1977-1981 only a limited number of these dwellings were made available to the elderly. If in the future too through-flow policy with respect to housing is not implemented (mainly the job of municipalities), the development of the shortage of dwellings for the elderly will depend mainly on the number of **newly constructed** suitable dwellings.

In addition to the more 'traditional' forms of housing for the elderly, in recent years experiments are being made with 'alternative' forms of housing. For instance some older people are found to show a preference for 'group housing'. It is however difficult to predict whether such an alternative can be counted on to play a role of any significance among the older population in view of the changes in concepts and attitudes which this involves (by comparison with the norms and values of the traditional nuclear family).[95)]
Moreover, it might be asked whether this form of housing would not be restricted to healthy older people, since providing assistance to less mobile members of the unit might constitute a too heavy burden on the other residents.
Secondly, there are the so-called kangaroo dwellings (that is to say the architecturally integrated combination of a nuclear family

dwelling and a dwelling for the elderly members of the family in one residential unit). As far as we know, experiments of this nature have only been carried out on a limited scale in the Netherlands. It would appear that success with this residential form will only be achieved if such dwellings are realized in the rental sector.[96]
In general, up to the present, studies as to dwelling forms have been of the nature of residential preference studies. Residential preferences - also those of the elderly - will always be extremely heterogeneous. One problem encountered is however that in such a survey the elderly themselves proceed from a downhill residential trend. With increasing invalidity, they see no alternative other than the usual: to move to a smaller more suitable dwelling, next to a residence for the elderly, and then, via a home for the elderly, to a nursing home. Especially the last two are viewed as a 'necessary evil'. In general the elderly do not envisage an alteration of exising solutions.

The ideas and experiments developed are usually put forward by 'experts'. There are warnings against introducing changes in policy with respect to housing over the heads over the elderly. The elderly themselves should be more involved in the designing of policy. To this end, a further-reaching process of emancipation would be necessary, which would also make the elderly more recognizable as a specific group.[97]

Further study is required as to various aspects of housing for the elderly. A list of questions for study is for instance to be found in a literature study by Serail and in the Working Document Nr. 1. 'Residences and the Elderly'[98] of the Steering Group for Research on Ageing.

2.5.12 Co-ordinated work for the elderly

Co-ordinated Work for the Elderly (GBW) has been functioning since 1969. Initially, emphasis was especially on the co-ordinating function of service centres, while at present the emphasis is mainly on the co-ordination of activities of various facilities. The Memorandum Flanking Policy for the Elderly and the Note on Policy Relating to the Elderly (1983) have given the GBW a new perspective by according it an important position in the so-called 'flanking policy'. The goal of flanking policy is not to create new facilities of any kind for the elderly, but rather to ensure that the functions

of the relevant facilities for the elderly are put to better use. In this context the emphasis is on supporting the initiatives of the elderly themselves and of those in their immediate surroundings, and of volunteers.[99] A distinction is made between flanking policy in the narrow sense and in a wider sense. In the first form of flanking policy, the emphasis is on facilities, with regard to maintaining the independence of the elderly. On the one hand this encompasses the district function of homes for the elderly, and on the other hand the co-ordination of intra and extramural facilities with a view to furthering efficient functioning and achieving maximum attunement of formal and informal help. The second form of flanking policy focuses on renewal, the emphasis being not on the care sector, but on technological innovation. These matters often fall outside the strict limits of policy for the elderly, such as the development of apparatus, goods and services for introduction in dwellings, communication, public transport, recreation, etc.
These innovations would serve to further possibilities for the elderly to continue to reside independently and to participate independently in social life outside of the home.[100]

The GBW forms part of the flanking policy in a narrow sense. In view of overlapping of tasks, the position of the GBW will have to be carefully assessed. In the reference scenario it is assumed that till the year 2000 the GBW will become increasingly important as an aspect of flanking policy.
Efforts to achieve a cohesive and integrated level of facilities in which the needs and wishes of the elderly are met as far as possible must in the first place originate at the local and regional level.

2.6 The economic context

For some years already efforts have been made to control costs in the area of health care. Total costs have risen from nearly 11 milliard guilders in 1972 to more than 32 milliard in 1982. The main reason for the limited success in restricting costs up to the present stems from the manner of financing via the insurance system in the area of health care.[101] Attempts to control costs by the fixing of tariffs and the lowering of subsidies did not result in the envisaged lessening of expenditure in health care. Also as a consequence of this explosive development of costs, various facets of health care came under discussion. Whereas in the early seventies the main concern was how better counselling, preventative mass examinations, and early intervention in certain disease processes could keep people more healthy or at least less ill, at present the main question is where the limits of care must be set, and how within these limits priorities can be determined.

This section deals with macro-economic developments till the year 2000 and the influence they exert on health care and services: in how far does economic growth or shrinkage affect health care and services? In our opinion, this question cannot be seen as apart from the above discussion as to the limits of care. What percentage of the national income are we prepared to devote to health care? How will priorities with respect to care develop?

Making forecasts as to long-term economic developments is a tricky business. We will consequently not attempt to do so. By restricting the question, however, somewhat more concrete pronouncements are possible. For instance, the Social and Cultural Planning Bureau has calculated what degree of growth of real national income will be necessary to be able to meet the growth of collective expenditure resulting from demographic developments. This amounts to a so-called null variant: point of departure is the actual collective expenditure in 1981. It has been calculated how great would be collective expenditure in the sectors 'social security', 'education', 'health care' and 'services' in the period 1981-2030 if a constant age-specific usage is assumed. 'Other sectors' are bundled in a fifth category. The results are given in Table 2.27. Estimates are based on the low variant of the population forecast.[102)]

Table 2.27 Development of total collective expenditure, 1981-2030, according to low variant of the CBS population forecast

year	1981 amount (mrd.gld)	1981 index[a]	1990	2000	2010	2020	2030
social security	56.3	100	111	121	131	139	140
education	20.2	100	85	80	77	67	63
health care	15.8	100	111	118	123	129	134
services	5.4	100	104	112	120	129	140
other sectors	100.5	100	104	108	107	104	100
total	198.2	100	105	110	112	113	112

a = index figures with 1981 = 100.
Source: Goudriaan et al., 1984, pp. 9-10

Table 2.28 shows the development of total collective expenditure for the three variants of the CBS forecast.

Table 2.28 Development of total collective expenditure according to low, medium and high variants of the CBS forecast, 1981-2030 (a)

year	1981	1990	2000	2010	2020	2030
Low variant	100	105	110	112	113	112
Medium variant	100	105	112	116	118	118
High variant	100	106	114	120	124	126

(a) = index figures with 1981 = 100
Source: Goudriaan et al., 1984, p. 11

In the period 1981-2000 collective expenditure increases by 12% in the medium variant. This implies 0.6% per annum. The real national income will also have to increase by 0.6% to meet the growth of collective expenditure resulting from demographic developments.[103)] This percentage is the result of the increasing proportion of older people in the population and decreasing number of younger people (compare the decline of expenditure in the educational sector and the increase in other sectors). The above calculations give only a rough picture. A large number of simplifications have been introduced in the calculation model.[104)] Considerable financial shifts within and between the various sub-sectors of the collective sector will be necessary as a result of the changes in the composition of the Dutch population according to age, civil status and sex.[105)] This is brought about not only by shifts in the size of the various age groups, but also because of the large differences in expenditure accruing to the various age groups. See Table 2.29 on this matter.

The total collective expenditure per elderly person (65+) is on the average two and a half times as high as the collective expenditure for younger people.

See Table 1.1 in Chapter 1 for a synopsis of the reference scenario.

Table 2.29 Distribution of persons and collective expenditure 1981 and average collective expenditure per person, 1981, five age categories

age categories	0-19	20-44	45-64	65-79	80+	total
- persons (x1000)	4363	5397	2831	1326	329	14246
- persons (in % of the total)	31	38	20	9	2	100
- collective expenditure (mrd. gld.)	60.0	57.1	40.2	30.2	10.7	198.2
- collective expenditure (in % of the total)	30	29	20	15	5	100
- average collective expenditure per person (in guilders)	13.760	10.580	14.190	22.760	32.430	

Source: Goudriaan et al., 1984, pp. 172 and 174.

Notes Chapter 2

(1) Research Group Planning and Policy-making, 1981, p. 211.

(2) CBS, 1982.

(3) CBS, 1982/5, pp. 6-7.

(4) It should be noted that recent data as to the relative share of the elderly in the population, in which the forecasted developments for 1981 and 1982 are compared with the actually observed developments, indicate that the **low** variant of the CBS is nearest to reality: that is to say, the **relative** share of the elderly in the Dutch population will increase. This however exerts little influence on the **absolute** share.

(5) The term **double ageing** is applied to the phenomenon of an increasing share of very old people in the category of the elderly, while the share of elderly in the total population also increases.

(6) It is not always possible to draw clear lines between the terms used here. The literature too gives diverging definitions. This is partly to be attributed to the complexity and intertwinement of the phenomena. For instance, invalidity can be caused by illness, but may also result from accident. Moreover, there are many forms of invalidity which also cannot always be clearly demarkated: psychical invalidity, mental invalidity, and, especially among the elderly, invalidity with respect to Activities of Daily Life. This ADL invalidity may in turn result from physical or mental disturbances.

(7) Inter alia Janerich, 1984.

(8) Quoted in the report of the Social and Cultural Planning Bureau, 1984, p. 29.

(9) CBS, 1982, p. 38.

(10) Van der Maas, 1982, p. 74.

(11) Hoogendoorn has calculated that the number of deaths will increase by 32.3% if the mortality pattern of 1974 is applied to the population structure of 2000 (population forecast 1976, alternative B). See Hoogendoorn, 1977. The influence of the **absolute** growth of the population has not been taken into consideration.

Mortality according to groups of causes, 1975

cause of death	registered in 1975	calculated on basis of popul. comp. per 1.1.2000	increase in %
Diabetes mellitus	1,576	2,149	+ 36.4
Neoplasms	29,433	37,990	+ 29.1
Coronary diseases	34,246	46,468	+ 35.7
Cerebrovascular lesions	12,663	17,536	+ 38.5
Pneumonias	2,156	3,083	+ 43.0
Bronchitis/emphysema/ asthma	3,759	4,841	+ 28.8
Congenital defects	850	690	- 18.8
Certain causes of perinatal mortality	909	645	- 29.0
Traffic accidents	2,437	2,527	+ 3.7
Other accidents	2,923	3,990	+ 36.5
Other diseases	22,785	30,543	+ 34.0
All causes of death	113,737	150,462	+ 32.3

(12) Cf. the views of the Royal Dutch Society of Medicine in Medical Contact, August 1984.

(13) Hollander, 1984a.

(14) Multimorbidity varies from an average of slightly more than 3 medical complaints per elderly person examined by a GP to as many as 6 complaints in a geriatric institution (Fuldauer, 1968, p. 99).

(15) CBS, 1984.

(16) When speaking of residents of homes for the elderly, we do not take into account those in sick-bay and monitored dwellings. The fact that persons in sick-bay are excluded could result in a distorted picture.

(17) CBS, 1976, p. 14.

(18) ADL = Activities of Daily Life, namely the following eight activities:
- eating and drinking
- sitting down and getting up out of a chair
- getting in and out of bed
- getting dressed, inclusive of putting on shoes and undressing
- getting to another room on the same floor
- going outside the house
- washing face and hands
- washing completely (take a bath or shower).

This is the CBS interpretation of the ADL concept. In 1982 the CBS added two more activities to the ADL scale:
- climbing and descending stairs
- getting in or out of one's residence.

(19) Van den Hoogen et al., 1982. Use is made of the population forecast 1980 of the CBS (variant unknown) and the data of four GP practices affiliated to the Continuous Morbidity Registration (12,000 persons).
In how far these practices can be considered representative for the Netherlands is not complete clear (variation in GP variables, patient variables, characteristics of the practice, etc.).

(20) These are changes resulting from the altered composition of the population **as a whole.**

(21) CBS, 1984, p. 10.

(22) Havighurst, 1978, p. 43.

(23) Swedish Ministry of Health and Social Affairs, 1982, p. 54.

(24) From: Rights and Regulations of the COSBO-Netherlands, February, 1984.

(25) The Association COSBO-Netherlands includes three member organizations, namely (1) the Association of Catholic Unions of Elderly (Unie KBO), (2) the General Dutch Union of Elderly (ANBO), and (3) the Protestant Union of the Elderly (PCOB). In total, COSBO-Netherlands has approximately 400,000 members. COSBO-Netherlands has a working agreement with the LOBB (National Organization of Committees of Residents of Homes for the Elderly).

(26) See inter alia COSBO, 1983.

(27) Havighurst, 1978, p. 38.

(28) CBS, 1982, p. 30.

(29) SCP, 1984, p. 369.

(30) SCP, 1984, p. 360.

(31) It is sometimes argued that the relation between the (increasing) emancipation of women and the (decreasing) willingness to provide umbrella care is not borne out by studies in other countries. For instance it is pointed out that a higher percentage of women have jobs outside of the

home in other countries, though at the same time, the percentage of elderly persons who reside **outside** of institutions is higher than in the Netherlands.
The following considerations should however be taken into account when assessing the relevance of this research material for the relationship between emancipation and umbrella care:

- in the above, not being institutionalized is equated with receiving umbrella care from women. It is however quite possible that non-institutionalized older persons in other countries do not so much receive umbrella care but rather formal ambulatory care from **professional** providers of aid.
- the relationship between emancipation and umbrella care could possibly be better assessed **per country** since the **entire** social and health care structure varies per country.

(32) Hilhorst and Verhoef, 1979.

(33) When the object of medical intervention is to end life, this is referred to as active euthanasia.
Where medical intervention is intended not to end life, but to reduce suffering, even though the drugs employed will undoubtedly hasten death, this is referred to as passive euthanasia.

(34) In how far the gauge stations affiliated to the CMR are representative for the Netherlands as a whole, is unclear.

(35) For a description of the various theories see Braam, Coolen and Naafs, 1981, p. 204.

(36) These publications include:

- The Final Report of the Committee for the Harmonization of Estimates for the Quaternary Sector plus the Technical Report (September 1983);
- Note on Policy for the Elderly (June 1983);
- Collective Expenditure and Demographic Developments 1970-2030 (SCP Working Paper No. 38, 1984);
- Costs and Financing of Facilities for the Elderly 1981-1991 (SCP Working Paper No. 39, 1984).

Projections with respect to one specific facility have meanwhile appeared, for instance 'An Estimate of the Work of Cross Societies for the Years 1985-1988' of the National Association of Cross Organizations, published in February 1984. In addition to this, there are a number of older reports. 'Looking after the Future' (SCP Working Paper No. 26 dating from 1981) is an example, but also the memorandum 'Aspects of Policy Relating to the Elderly' (dating from 1982) contains projections of the number of places that will be required in nursing homes in 1990.

(37) Moreover, in some regions a saturation point has been reached for facilities (for the elderly), namely in Friesland and Limburg.

(38) Pommer and Wiebrens, 1984, p. 44.

(39) SCP, 1984, p. 19.

(40) CBS, 1984/10, p. 7.

(41) Fuldauer, 1973.

(42) See for instance Van Genuchten, 1983, GP's problems with psychogeriatric patients.

(43) Data of the Dutch Institute of General Practitioners. In January 1983 the total number of independently established GPs was 5616.

(44) This possibly refers to GPs who start a joint practice to facilitate the transition of the practice to a successor.

(45) Memorandum on First Echelon Care, p. 17.

(46) Willemse, 1981 and Brouwer, 1983.

(47) NVAGG/GHIGV, 1979, p. 3. This does not include mentally disturbed people who have grown old, or older mentally retarded people. These fall under the social-psychiatric function within the AGGZ.

(48) This refers to all patients of 65 years of age and older. The use of psychogeriatric aid cannot be seen as distinct from other forms of aid. The age of 4361 patients (2.8%) was unknown. It is also not known how great was the non-response in 1982 to the annual survey as to the number of patients of affiliated institutions.

(49) Goudriaan, De Groot et al. 1984, p. 103.

(50) Ten Horn, 1981.

(51) Memorandum on First Echelon Care, p. 28.

(52) State Secretary for Welfare, Health and Culture, 1983, p. 26.

(53) Statistics of homes for the elderly 1979 and 1980. The Hague, 1983 (CBS).

(54) Pommer and Wiebrens, 1984, p. 45.

(55) The authors of this SCP report base their calculations on those of the Committee for Harmonization of Estimates for the Quaternary Sector. See 'On the Use of Facilitaties and Staff in the Quaternary Sector 1983-1987'. Technical report accompanying the final report of the Committee for Harmonization of Estimates for the Quaternary Sector, Ministry for the Interior/SCP, The Hague, September 1983.

(56) Report on a study as to the sharpening of indication policy for admission to homes for the elderly in the province of Utrecht carried out by DHV Consultation Bureau Ltd., on the commission of the Provincial Council for Public Health of the province of Utrecht, Amersfoort 1984. The average age of those who had been found eligible for admission in 1982 and 1983 was 82 years.

(57) DHV, 1984.

(58) This is preferred since the elderly remain relatively long in homes for the elderly (an average of slightly more than 6 years), and the influx in the period 1975-1978 dropped radically. In this flow model designed by the HARK Committee 16 categories are distinguished (married/unmarried, male/female, and four age categories). On the basis of the data on the end outflow of the elderly in the period 1975-1981, the future number of residents are calculated. The HARK Committee calculated this till 1987 and the SCP till 1991 (Pommer and Wiebrens, 1984).
In the latest estimates concerning the quaternary sector (Goudriaan, De Groot et al., 1984, p. 111) a highly simplified flow model is however employed, based on only two age groups: 65-79 year olds and 80 year olds and older. The results of this simplified flow model were found hardly to deviate from the results of the more complicated model (less than 1%). Since 1978 the in and outflow of elderly people in homes per age category has remained almost constant, as a consequence of which estimates in the trend variant coincide with the null variant. Results deviate from the earlier estimates (cf. Table 2.18). This is brought about partly by the application of the low variant of the population forecast instead of the medium variant. (1983: 139,300; 1985: 142,300; and 1990; 153,200).

(59) Though the null variant assumes constant age and sex-specific usage figures per five-year cohort, the number of people in care institutions as a percentage of the whole group of people of 65 years of age and older is on the increase. This is a consequence of the increasing share of very old people within the group of persons of 65 years of age and older.

(60) With respect to all facilities no estimates are given in the trend variant for the year 2000, since it is not statistically justifiable to make estimates further into the future than the number of years that have served as base period for calculation of the trend. The base period employed by Pommer and Wiebrens is approximately 10 years for the various facilities that will be presented here.

(61) Minister of Welfare, Health and Culture et al., 1983, pp. 23 and 24.

(62) Pommer and Wiebrens, 1984, p. 35, and Goudriaan et al., 1984, p. 113.

(63) Year-Book of the Nursing Home Information System (SIVIS), 1982, p. 75.

(64) Minister of Welfare, Health and Culture et al., 1983, p. 21.

(65) This conclusion is somewhat artificial since during sojourn somatic patients are more often placed in psychogeriatric wards than the other way around. Year-book (SIVIS), 1982, p. 41).
A further complicating factor is that the average stay in a nursing home is also heavily dependent on the date when the facility was opened.

(66) National Association of Cross Organizations, 1984, p. 33.

(67) Bartels, 1984.

(68) State Secretary for Welfare, Health and Culture, 1983, p. 27. This study has recently been reported on. See Nuy, et al., 1984.

(69) Hospitals Institute of the Netherlands, 1981.

(70) Van Santvoort, 1984, p. 26.

(71) Pommer and Wiebrens, 1984, p. 48 ff.

(72) The norm applied to general and special hospitals.

(73) State Secretary for Welfare, Health and Culture, 1983, p. 31.

(74) Ministry for the Interior/SCP, 1983, p. 128.

(75) No sex and age-specific estimate.

(76) National Association of Cross Organizations, 1984, p. 31.

(77) From an article in the monthly magazine Social Health Care, it was found that only 11 basic units are involved, and that this number is too small to give a representative picture. See Bakker-Lenderink, 1983, p. 9.

(78) Is is unclear whether this proceeds from the increase of **independently residing** elderly people or of all the elderly.

(79) Age-specific calculation.

(80) Pommer and Wiebrens, 1984, p. 44 ff.

(81) Pommer and Wiebrens, 1984, pp. 49-51.

(82) National Association of Cross Organizations, 1984, pp. 31-32.

(83) National Association of Cross Organizations, 1984, p. 11.

(84) Note on Policy Relating to the Elderly 1983, p. 20. In the Memorandum on Health Policy With Limited Resources, this growth in volume is also set at 4% for the period 1984-1986.

(85) Memorandum on First Echelon Care.

(86) Ministry for the Interior/SCP, 1983, p. 178. The drop is connected with the fact that the budget was exceeded in 1976, and the consequent correction in later years. There is a significant shift in trends, which is brought about especially by the sharp drop in the amount of aid to people under 55 years of age. Use of home help by the elderly shows a declining growth.

(87) CBS, 1979, pp. 61-62, and Mootz and Timmermans, 1981, pp. 71-72.

(88) No single nomenclature exists. See for instance NFB, 1974.

(89) House-boat, temporary dwelling, etc.

(90) This relates both to those who already reside in a suitable dwelling for the elderly, and those who would like to move to such a dwelling.

(91) Rongen and Houben, 1984, p. 47.

(92) Rongen and Houben, 1984, p. 13.

(93) Minister of Welfare, Health and Culture et al., 1983, p. 14.

(94) Of the total number of suitable dwellings for the elderly in 1981 in the Netherlands (620,881), 36% were inhabited by households the head of which was younger than 55 years of age.

(95) Scientific Council for Government Policy, 1982, pp. 154-155. Research as to new residential forms was carried out inter alia by the RIW Institute for Housing Research in Delft, the Dutch Federation for Policy for the Elderly (NFB), the IVA Sociological Institute in Tilburg, and by Humanitas.

(96) When such housing units are realized in the purchase sector, they often end up in the hands of persons who use them as office quarters etc.

(97) Houben, 1983, pp. 488 and 491.

(98) Serail, 1982, pp. 9 ff. and Burie (ed.), 1984.

(99) Ministry of Welfare, Health and Culture, 1983, p. 9.

(100) Ministry of Welfare, Health and Culture, 1983, pp. 53-54.

(101) SCP, 1984, pp. 47-48. An insurance gives the right to certain treatments, as a consequence of which the number of treatments and the costs can never be determined.

(102) For the medium variant data are only available for total collective expenditure.

(103) Assuming a constant ratio between national income and collective expenditure.

(104) Goudriaan et al., 1984, pp. 168-172. The most important are:
- Constant usage of government programmes (constant usage figures according to age, sex, and civil status).
- The idea of having government programmes taken over by private initiative and introducing or increasing clients' contributions has been disregarded.
- It has been assumed that government programmes will remain unaltered.

- Unaltered expenditure per unit of use; in the past there was however a sharp increase in expenditure per unit, as a consequence of which future development of expenditure has probably been (heavily) underestimated.
- No conscious influencing of demographic developments (population policy).
- The per capita net national income (at market prices in stable guilders) remains at the level of 1981.

(105) In the year 2030 nearly double the amount expended in 1981 will be required for old-age pensions, while child allowances in 2030 will be only two-thirds of the amount for 1981 (Goudriaan et al., 1984, p. 181).

3 Medical and medical-technological developments

3.1 Introduction

The amazing strides in medicine and social facilities during the last century, manifested in better health care facilities, hygiene, vaccination, the introduction of antibiotics etc., have had the effect of doubling life expectancy at birth in the industrialized countries.[1)]

The first medical revolution is coming to a close. The present health situation and the system of care are characterized by relatively modest progress if developments are measured by health indicators (such as life expectancy, morbidity, etc.). There is hardly any further increase in life expectancy, there is no or hardly any **net** decrease in the prevalence and incidence figures of the most important diseases such as cardiovascular diseases and cancer, and there are still no prospects for large-scale cures of these diseases, while the need for care continues to increase. Moreover, the costs of health care and biological and pharmaceutical research have risen enormously. Given this combination of limited progress and high cost increases, it is interesting to study how this relationship will appear in the future, and what may be expected of medical sciency in general and biomedical research in particular.

What are the expectations with respect to breakthroughs in medicine, pharmacy and medical technology which will exert an influence on the state of health of the elderly till the year 2000? Are developments observable which could possible be the heralds of a new (second) medical revolution? And what are the possible consequences for the state of health and the volume and nature of care facilities?[2)]
To obtain answers to these questions, the scenario commission decided to augment its expertise by seeking the advice of thirty specialists deriving from various fields. Around May 1984 six consultations were organized with experts from the areas medical and medical-technological research and medical practice.
Participants were selected in close consultation and co-operation with the Royal Dutch Society of Medicine and the Steering Group for Research on Ageing. The consultations were intended to give a wider foundation to pronouncements in the scenarios on medical and

medical-technological developments between 1984 and 2000 than would be possible on the basis of information deriving from the literature and from knowlegde of members of the scenario committee. The participants were asked to check the reports of the consultations. On the basis of reactions, one final report was written. Together with the names of the participants, this is affixed to the scenario report as Appendix I.

The general conclusion is that till the year 2000, no real breakthroughs or developments are to be expected which would exert any fundamental influence on the present state of health and the system of health care. Neither are breakthroughs expected in the treatment and healing of diseases in the short period till the year 2000. Developments already in progress, especially in surgery, will benefit the elderly. In general these developments will not have the effect of prolonging life, but rather will help to combat invalidity and will improve the quality of life.

The following important reasons are quoted for the continued absence of breakthroughs in medical science.
In the first place, the span of 16 years is considered to be too short to realize changes, since some time is required for making fundamental breakthroughs in research applicable and for introducing them into medical practice. In the second place, many policy decisions which exert an influence on the future structure of health care are already being taken, and will apply to medium and long-term developments.[3] Lastly, it will hardly be possible to exert any influence on the health situation of this and the next generation of elderly, since most of the diseases from which people suffer at that age follow on long latent periods. In other words, the diseases are already in an advanced stage, so that preventative measures are no longer possible. This leads to the conclusion that, irrespective of (unlikely) breakthroughs, the state of health of the present and future generation of elderly till the year 2000 will not alter significantly, and that the picture of the state of health of the group of elderly as a whole will be mainly determined by the phenomenon of double ageing.
Predictions for after the year 2000 are of a different nature. In spite of the fact that the period is too far distant to make reliable pronouncements as to breakthroughs which might come about in health, there is nonetheless a reasonable consensus of opinion as to possible future changes and developments in fundamental research. It is thus expected that there will be changes in the state of health of the elderly of after 2000.

In view of the absence of fundamental breakthroughs in medical science, and the consensus of opinion on this matter among external experts, the committee has decided not to formulate alternatives for the variables 'medical and medical-technological developments' per scenario, but to treat them as one context.

Developments and breakthroughs will be discussed in the following sections. Where possible, breakthroughs after 2000 will be indicated and discussed, but will be treated as 'Low probability-High impact' events for the scenarios, since they are expected to come about only after the year 2000.

3.2 Expectations with respect to medicine and pharmacology

For the immediate future, diseases with a long latent period (including some types of cancer, coronary and arterial diseases, dementia and diseases caused by persistent viruses) will continue to play an important role in the state of health of the population.[4]
In the event of extension of the average life expectancy and improved possibilities for treatment of diseases not known, it is expected that 'new diseases' with an even longer latency period will make their appearance. The great disadvantage of existing diseases with long latency periods is that the effects on the range of disease in the coming generation(s) will change little; in other words, if no drugs are found to cure these diseases, and assuming a similar lifestyle and dietary pattern in the coming generation, there will be no change in the prevalence of disease by comparison with the former generation. For the future this means that the prevalence of disease will increase in an absolute sense as a consequence of the ageing of the population. In coming generations there might be a shift in the prevalence and prevention of disease if there are changes in dietary patterns and lifestyles (compare the present popularity among the young of fast foods, increased obesity, substitution of dietary calories by alcohol calories, the decline in smoking).
It is expected that changes in the lifestyles of younger generations will bring about changes in health, as a result of which relative changes per disease might occur.

As far as cure or retardation of disease processes is concerned, it is not expected that there will be any breakthroughs with respect to the most important diseases. No important changes in the present research models are expected in the coming 20 years which would result in any significant breakthroughs in the treatment of diseases. In as far as there are any breakthroughs, these will be in complementary disciplines such as immunology, genetics, pharmacology and the application of new examination apparatus.[5] It is expected that only very modest progress will be achieved with respect to the various diseases. For the future, the emphasis will be mainly on palliative measures and the ensuing improvements in the quality of the life of the patient. Possible breakthroughs may only be expected in prevention (that is to say, either preventative measures or changes in personal lifestyle). Possibilities for altering lifestyles should however not be overestimated. Cancer of the lung and cirrhosis of the liver are for instance two diseases which are largely brought about by lifestyle.[6] In addition to prevention, expectations focus strongly on early diagnosis of those diseases which react to treatment. One should however be on one's guard against discrepancies between diagnosis and therapy. The early diagnosis of diseases for which there is no therapy has rather a negative than a positive effect.

Multimorbidity with possibly mental deterioration plays an important role among the elderly.
As a result of medical and medical-technological innovations, the elderly can benefit at an increasingly later age from medical intervention. Especially through corrective measures and the treatment of traumas, surgery is in a position to maintain or to improve the ability of the elderly to remain active. Moreover the advent of a new specialism is expected, namely that covering the pelvic region. The existing specialisms of urology, surgery, neurology and gynaecology would unite their forces in this new specialism with a view to improving treatment of the diseases and complaints in this region among the elderly.
As a consequence of the often encountered multimorbidity and lessening of mental capacities among the elderly, the physician is increasingly obliged to decide against treatment, in spite of improved methods of treatment.

Expectations with respect to the effect of developments in pharmacology for curing diseases are limited. On the one hand, the pharmaceutical industry is dependent on insights deriving from medical science with respect to breakthroughs in the area of etiology.
For the time being, the lack of fundamental knowledge and insight into most disease processes, and the expectation that there will be no change in this situation in the coming 15 years, leads to the

conclusion that no breakthroughs may be expected in the area of pharmacology. Present research strategies may yield fruit, but chance plays such a large role, that is is impossible to make any predictions as to developments. The situation would be completely altered if disease processes were better understood. It would then become possible to design medicines by means of a far cheaper research strategy.
On the other hand, the pharmaceutical industry is dependent on economic remunerativeness and will consequently not focus on every disease, but only on those which can count on a sufficiently large market to make production worthwile. Rare and complicated diseases will consequently receive little attention. Government could play a role in stimulating (financially) risky pharmacological research in these areas.

3.3 Expectations with respect to technology

3.3.1 The costs aspect

Medical-technical apparatus serves to aid the physician in making a diagnosis and in implementing and monitoring therapy.
Apparatus becomes qualitatively increasingly better, but also increasingly complicated and expensive. Take for instance introduction of the NMR. Not only is the apparatus itself costly, but it is also costly to install and to use. On the other hand, there is also a tendency to make high-quality apparatus which is easier to use. This apparatus is mainly intended for use by a wider public.

The costs factor plays a crucial role for the future development of technological apparatus.[7] From a purely technical point of view, possibilities already exist for solving many practical problems, but only at high cost. Moreover the question arises in how far these relatively small improvements and refinements justify the costs involved in introducing this apparatus. As against this, once the apparatus has been introduced, prices may be reduced by production on a larger scale, while the disappearance of techniques applied earlier will also mean a reduction in costs.
Moreover, the introduction of (expensive) aids may have the effect of reducing the costs of care, since they may aid the elderly to function independently for a longer period of time.

3.3.2 Medical-technical apparatus

No fundamental breakthroughs are expected in the improvement of therapeutic apparatus. Improvements are however expected in scanning techniques for making diagnoses and for monitoring of therapy. Examples of this are the PET scan, NMR and three-dimensional echography. In general these new scanning techniques do not result in better diagnoses than are obtained with the existing scanning techniques. The advantage of the new generation of scanners is especially the possibility for obtaining more extensive diagnostic data, which can be used in fundamental research as to the origin and course of disease processes and the effects of therapy.
In addition to scanners, lasers will assume an important role in heart and eye operations (microsurgery).

3.3.3 Technical aids

As far as the development of technical aids is concerned, it is expected that there will be vast strides in innovations intended for the elderly through application of microelectronics. These aids often do not serve to prolong life, but considerably improve the quality of life of the (handicapped) patient and make it possible for him to function longer independently and to remain in his usual surroundings.
In this context, considerable progress is anticipated with respect to prostheses. It is expected that in future, in addition to prostheses of the knee and hip, prostheses of the hand and foot will be introduced. In particular expectations are high as regards 'intelligent' prostheses. These self-operating prostheses are easy to use, since by means of biofeedback system, they function independently of the knowledge of the user. One prerequisite for the application of this type of prosthesis is a well-functioning nervous system. Multimorbidity with possible mental deterioration considerably limits possibilities for applying them in the case of the elderly.

Another problem which needs to be solved in the case of intelligent prostheses (especially the artificial eye and artificial ear) is connecting the prosthesis to the correct nerves. In the case of younger patients this presents less of a problem since the plasticity of the brain is many times as great as in the case of older people. In the young, it is much easier for brain functions to be taken over by other parts of the brain than in the case of the elderly. A possible solution to this problem is the presence of 'reserve receptors' distributed through the body, the purpose of which has not yet become clear. Often it is possible to make use of these reserve receptors instead of the malfunctioning receptors (an instance of this is 'seeing with the stomach'. Here the artificial eye is connected to reserve receptors in the stomach which can distinguish between light and dark).

In addition to prostheses, technical aids will be developed for making it easier for the older invalid individual to perform daily activities, so that he can continue for longer to function independently in his own surroundings. In this connection comes to mind the production of vehicles adapted to the elderly. If produced en masse, this would reduce production costs.
One problem with this new apparatus is that is becomes increasingly compact and complicated, which can result in problems with operation, especially among the elderly (imagine for instance a person with a tremor), apart from the possible aversion of elderly people to complicated apparatus, the functioning of which they do not completely understand. A counter-trend is however observable for making this high-quality apparatus easier to operate.

In addition to the above, much is expected of implanted aids. We have in mind drug reservoirs which, once swallowed (in tablet form) or implanted (insuline pump), dispense uniform doses of the drug and continue to function for several days. Furthermore improvements are expected in methods of fuelling the artificial heart, as a result of which it will be possible to employ it more often (in addition to heart-lung transplantations). Electronic heart regulation (pacemaker) is still very much in the process of development, and in this area too intelligent apparatus is expected which will be much smaller and generally improved. The same is expected with respect to artificial sphincter muscles for the solution of incontinence among the elderly with a well-functioning nervous system.

In addition to the above, vast strides are expected with functional polymers, especially organic-chemical substances for transplantation and growth stimulation (jaw prostheses and reservoir pills).

3.3.4 Nursing

In the coming period till the year 2000, developments in nursing may be expected which will benefit the elderly. Nursing of the hospital patient in bed will make way for nursing out of bed. By this means, a number of now common complications among the elderly could be avoided, such as decubitus, contractions, infections of the urinary tract, incontinence of urine and faeces, pneumonias, mental deterioration and mental confusion. There will be a shift over from passive nursing to nursing focused on activiation of the patient with a view to maintaining both the psychical and mental condition at the highest possible level. To this end, attention will be devoted to the development and application of ergonomic principles in nursing.

3.3.5 Information technology and epidemiology

As a result of the enormous possibilities for application of micro-computers, it is expected that there will be vast improvements in information technology: for the patient as a tool for information and communication; for the physician as an aid and check in the making of diagnosis.
Epidemiology is also expected to benefit from this trend as a consequence of possibilities for collecting and processing large quantities of medical information. By this means insight can be obtained into the causes and distribution of diseases as well as the risk factors attaching to the diseases.
We should however not fail to draw attention to the problem of protecting privacy, an aspect which should not be underestimated. Pleas are heard for regulations for the adequate protection of the privacy of patients which will at the same time not lead to unnecessary slowing down of epidemiological research.[8)]

3.4 Towards a second medical revolution?[9)]

In some medical sectors not only a stagnation of developments, but even a slight reversal in developments is observable. In this section, an attempt is made to explain this phenomenon against the background of the research strategy adhered to in medicine and pharmacology, while also endeavouring, on the basis of possible future breakthroughs in fundamental research, to formulate expectations for future developments in the two areas.

From the start, research strategy in medicine and pharmacology proceeded from the need to understand and control disease processes. Due to an often insufficient knowledge of cells and cell processes, emphasis was often placed more on the control of disease processes than on the understanding of them. The living cell and the organism was an as yet unopened black box. What was wanting in actual insight into processes was compensated for by means of trial and error. In this period the chance factor (the serendipity component of research) played a not unimportant role in the solution of certain problems.
Considering the enormous improvements in the state of health of the population, this research strategy proves to have been highly successful. Present progress however leaves much to be desired; the prevalence of important diseases does not decrease or even increases. Even where causes are known, in practice it is often found impossible to remove them due to a combination of technical, social and economic factors.
This stagnation is however not of a permanent nature. Thanks to a veritable revolution in molecular biology brought about by new research methods and techniques, great strides have been made in understanding the structure and function of living cells, and discovering how energy is obtained, labour performed, and hereditary traits reproduced. As a result of this vastly increased knowledge of the normal cell and the healthy organism, it has become better possible to describe pathological processes, to distinguish them and especially to understand them. Medical science is now in a position to adopt a more rational strategy which, proceeding from a knowledge of the normal cell, would make it possible to understand pathological forms and to design therapy. The differences between the old research strategies and the new are clear: on the one hand, the cell as a black box, the importance of chance and not completely understanding causes, as opposed to the cell as point of departure for research, the designing of the correct therapy on the basis of knowledge of causes at the level of the cell. It is expected that after 2000 the fundamental knowledge of cells, heredity et cetera, will be available and that the fundamental change in research strategy will lead to a second medical revolution.

The section which follows discusses some fundamental studies in cell biology without going too deeply into all aspects. In the context of this report it would not be possible to discuss all future developments in cell biology.

3.5 Expectations with respect to cell biology

3.5.1 What is ageing?

In addition to research as to the causes and combating of specific diseases, fundamental research as to the process of ageing at the level of the cell plays an important role. In the course of ageing, a natural lessening and/or change comes about in cell functions in the body, as a result of which the chance of acquiring diseases increases. Or, as Makinodan defines it: 'Ageing can be defined as a time-dependent process whereby one's body can no longer cope with environmental stress and change as easily as it once could. Hence loss of physiological adaptability is one of the hallmarks of ageing'.[10)]

Improved knowledge of how this process is brought about, and when it begins for specific cells, makes it possible to bring an end to the present situation in which all sorts of theories as to ageing compete with each other.

When it is clear what exactly is ageing, and the factors behind the process are disclosed, it will be possible to distinguish between normal and pathological ageing. It is even conceivable that all sorts of symptoms which are now attributed to the normal process of ageing will be discovered to be pathological.[11)] With the possible new design strategy (Section 3.4), it will be possible to design all sorts of drugs which could cure or prevent these pathological forms of ageing. Possibilities would then perhaps be opened up for retardation of the ageing process or postponing it till a later age. One thus progresses from the curing of disease to the prolonging of life, to the improvement of the quality of life.
However, we have not yet arrived at this stage. Obstacles are still encountered in research as to the process of ageing at the level of the cell. There are for instance adequate animal models for studying the process of ageing of cells of the lower functions of the body, which can assist the researcher in interpreting the problem. When studying the process of ageing of higher functions as in the case of the brain however, no adequate animal models are as yet available. It is expected that around the year 2000 and in the years following, solutions will be found to this problem also.[12)]

3.5.2 Biotechnology[13)]

As was pointed out, hereditary information in the normal cell plays an important role in the ageing of the cell. This information is stored in coded form in the DNA, which forms part of the chromosomes of all cells. DNA molecules are built up of four chemicals, connected to form threadlike spirals. The coded information depends on the specific order of sequence of the chemicals. The genes represent the hereditary information which determines all characteristics of an organism. Mutations can come about if the sequence of the DNA is altered or by removal or introduction of parts of the DNA. Once a change has come about, this remains permanently in the genetic information of the cell. Mutations may come about spontaneously during replication, or may be caused by chemicals or radiation.
For instance, there are already considerable indications that changes of normal cells into cancer cells are brought about by a process similar to a mutation, which causes the cells no longer to respond to the body's usual growth regulation mechanisms.

When sequences of different DNA molecules are linked together, this is called a recombination. Recombination occurs within various chromosomes in the same cell with the result that a completely new sequence and grouping of genes comes about in the offspring. Till recently, it was extremely difficult and time-consuming to isolate genes of humans, animals or plants. For this reason, it was not possible to study the molecular structure of the separate genes, or to determine how they were ordered on the chromosomes. In the early seventies a fairly simple and rapid method was discovered by which means it has become possible to isolate the genes of nearly all living organisms, and to obtain them in nearly pure form and in practically unlimited quantities. It is expected that even before the year 2000 the human DNA will have been charted.

These developments lead to great application possibilities for medicine and pharmacy.
An already known application of the recombinant DNA technique is molecular cloning, whereby certain parts of the human chromosome are isolated and introduced into a bacteria, which then produces unlimited amounts of the genetic material which has been introduced into it. By this means human insulin is already produced for diabetic patients, growth hormones for certain genetic growth disturbances, enzymes for the treatment of blood clots and occlusion of blood vessels.

More spectacular applications of these techniques are conceivable for the future. Especially the treatment of hereditary diseases by substitution of healthy genes obtained by molecular cloning for the defective ones is often mentioned in this context, as well as the prenatal screening of foetuses in utero for serious hereditary defects.
Nonetheless, so many pitfalls and unknown quantities still attach to the 'substitution' of genes as a therapeutic approach, that questions arise as to the realizability and desirability of such therapy in the case of certain genetic defects, even apart from possible risks.
However, there have been enormous breakthroughs in the area of molecular genetics in recent years, which would appear to create possibilities for influencing and changing (hereditary) characteristics of living cells for purposes determined by man. This offers enormous potentials, but also involves vast responsibilities.

3.5.3 Immunology

The immune system constitutes the body's most important defence mechanism against foreign matter which invades the body. It is known that during the process of ageing changes occur in this system, as a result of which the change of acquiring certain diseases is heightened. An important role is played in this process by substances which regulate the passage of information between cells. It is expected that it will become possible to isolate these substances in a pure form, which will open possibilities for regulating the immune system. To start with, this regulation will be a-specific, only aiming at pepping up certain weaknesses in the immune system. Later, it is expected that more specific regulation will become possible, inter alia for purposes of intervention in the case of autoagressive antibodies (antibodies directed against components of the individual's own body). One application of this might be the combating of autoimmune diseases. The significance of this finding for the elderly is however limited in view of the limited number with pathogenic autoantibodies.

In addition to this, expectations are high with respect to the application of knowledge of the immune system to other areas, such as for instance endocrinology. In this context comes to mind the possibility of linking chemicals to antibodies which in turn can be directed at receptors, or designing autoantibodies which can combat a too high concentration of a hormone.

Immune deficiency is thought not to play a role in the development of cancer, though it does play a role in the combating of cancer, namely in the form of immunotherapy which comes about by the exogenous administration of antibodies with toxicants aimed at specific tumors.

An important application of immunology is in seeking a solution to the problem of transplant rejection. The body is strongly inclined to reject any new foreign matter. If the secret of steering the immune system is solved, it may be possible to find a solution to the question of rejection. Possibilities may exist in the future for treating tissue or organs for transplantation in such a way that the immune system may not recognize them as foreign matter. Possibilities will then also exist for making animal tissue and organs suitable for transplantation, thus creating new opportunities for combating deficiencies among the elderly.[14)]

3.5.4 Neurobiology

Another area in which knowledge of immunology may be applied is neurobiology. In this discipline, research as to changes in neurons during the process of ageing and dementia plays an important role, for instance with the aid of immunocytochemically identified transmitter systems, and on the basis of biochemical markers (for instance measuring of changes in synaptic membranes, neurotransmitters or enzymes).
In this context, the relation between 'normal' and 'pathological' ageing, especially in as far as these relate to Alzheimer's senile dementia (SDAT), calls for attention.
It may be expected that in the coming years more insight will be obtained into the systems which are primarily subject to change during ageing and SDAT.
Such observation could be the key to early diagnosis and rational therapy of some forms of dementia.

3.5.5 Conclusion

Summarizing, it is to be expected that in the near future there will be enormous progress in the area of fundamental biological research, which will have radical implications for medical science, both as

regards ethics and application. The process of ageing as such and diseases resulting from pathological ageing will become recognizable and subject to intervention. Knowledge of the causes of these processes will increase predictability, on the one hand through prenatal screening for hereditary diseases, and on the other hand by determining risk factors of non-heriditary diseases through mapping of the individual genome.
Elements of these fundamental breakthroughs are already visible, as are some applications. It will however take till after the year 2000 before the store of information and experience have reached a sufficiently high level to bring about a second medical revolution. An important aspect of this development will be the ethics involved. Possibilities for altering hereditary information according to one's own insights constitute only one modest example.

Social aspects too warrant attention. It may be questioned in how far early screening results in increased or decreased social solidarity as this effects development possibilities for the individual and the social investment in high-risk members of a society.

3.6 Concluding remarks

Though no important breakthroughs are expected till the year 2000, the prospects for after 2000 are much more optimistic, especially in that further cell biological research is expected to yield some very positive results. This implies that a positive attitude to current research is certainly justified.

Notes Chapter 3

(1) Knook, 1984, p. 244.

(2) De Duve, 1983; Burg, 1983, pp. 1-3.

(3) Burg, 1983, p. 1.

(4) See Appendix I for expectations with respect to developments in numbers and possibilities for the curing of certain diseases.

(5) Burg, 1983, p. 4.

(6) Attention should be drawn to the efforts of the American National Cancer Insitute (NCI) to push back cancer, both through improved diagnosis and through changes in lifestyle. The goal is to reduce by 50% the number of deaths brought about by cancer by the year 2000:

- 15% by discouraging smoking. This would imply that in 1990 there would be a 50% reduction in smoking by comparison with 1980;
- 15% from raising the treatment of cancer at a national level to that of the best treatment which is now available;
- 5% as a result of changes in diet, such as less fats and more roughage; and
- 15% through the development of new and more effective methods for discovering metastases in an early stage (with the aid of monoclonal antibodies).

(7) Burg, 1983, pp. 9-10.

(8) See Van der Maas 1984 and Van der Maas, Habbema and Hayes 1981 for further discussion of the privacy question and the consequences for epidemiological research.

(9) De Duve, 1983.

(10) Makinodan, 1977, p. 1.

(11) Hollander, 1984, p. 233.

(12) Rigter, 1981, pp. 122-134 and Goedhard, 1981, pp. 135-140.

(13) Berg, 1983.

(14) Earth Year 2050, 1984.

4 Scenario B: Increasing growth in demand for facilities

4.1 Introduction

In general, this scenario proceeds from a growth in the demand for facilities by the elderly which is greater than might be expected on the basis of demographic developments. The reference scenario discusses what will be the consequences for the various facilities of the double ageing in projections till the year 2000 in the so-called null variant.

The reference scenario also distinguishes a trend variant: how will the demand for facilities develop in the future if account is taken of past trends in the use of facilities. It was found that with respect to all the facilities distinguished there is an absolute growht in demand. Relatively speaking however (the percentage of the elderly who make use of a certain facility or the number of hours of aid per person), there is a decrease in use of homes for the elderly and hospitals, while the relative use of care by GPs and district nursing remains constant. A relative increase is found with respect to the other facilities (nursing homes, day treatment and home help).

In the growth scenario it is assumed that the demand for facilities will be greater than might be expected on the basis of the trend variant.

The increasing growth in the demand for facilities is brought about in this scenario by a number of more or less consistent developments which may be summed up as:

(1) Values and norms applying to the elderly in society remain or become more discriminatory.
(2) A high tendency to consumerism with respect to (medical) care.[1]
(3) Continuing professionalization of care.
(4) Less solidarity between the generations (less care provided by children for their parents, less other informal aid).
(5) Less solidarity within the generation of the elderly (less reciprocal aid among the elderly).

These developments fit a society type in which achievement, work, productivity and attachment to material goods constitute central values (cf. Section 5.2.1).

Compared with the reference scenario, a number of variables however remain identical. For instance, the growth scenario also proceeds from the medium variant of the CBS population forecast. The gap in life expectancy between men and women will decrease slightly in accordance with the CBS forecast (see Section 2.3). A number of social developments too remain identical (see Table 1.1 of Chapter 1). The most important of these is that the educational level of the elderly also increases according to this scenario.
The elderly will become more assertive and will in general be interested in furthering their development even at a later age. The trend mentioned under (1) however exerts a negative influence on this tendency.
The developments referred to are mainly of an autonomous nature: they are hardly influenced by policy with respect to health.

This chapter is structured as follows: in Section 4.2 the developments mentioned under (1) to (5) are discussed. The consequences of these developments for facilities for the elderly are indicated in Section 4.3. The reader is referred to Table 1.1 in Chapter 1 for an overview of the contents of this scenario (the variables and the direction in which these undergo change).
Scenarios A and C are also included in this overview, so that a comparison is facilitated.
Finally, the economic context receives attention in Section 4.4.

4.2 Social developments as they affect the health situation

4.2.1 The elderly of the future

In the reference scenario in Section 2.4 various social developments were described, from which it may be deduced that in future the elderly will be better educated, will behave in a more emancipated fashion, and will constitute a segment of society with increasing political significance. There will also be an increase in the (relative) prosperity of this group, especially as a consequence of better pensions.
It may be assumed that a combination of these factors will have the

effect of raising the aspiration level of the elderly and making them more assertive.
These points of departure are also adhered to in scenarios A and C. Does this imply that in the future the elderly will be more independent[2] and socially better integrated? The answer to this question is less unequivocal than would appear at a first glance.

Since the seventies, government policy has been aimed at helping the elderly to maintain their independence for as long as possible. The elderly themselves subscribe to this goal (in as far as this finds explicit expression via unions of the elderly, for instance). This goal would thus further the wellbeing of the elderly. There are however arguments to the contrary.
The Scientific Council for Government Policy for instance points out that a policy that strongly propagates the value of individual independence could injure the wellbeing of the elderly, and might even heighten the demand for facilities.[3] There are limits to the pursuit of independence. For some, the process of ageing means that the physical condition, with the associated functions (walking, sight, hearing), deteriorates. This has the unavoidable effect of making these people more dependent on others. If revalidation is no longer possible, it is essential for the individual to accept this position of dependency if he is to experience ageing positively. If one refuses to accept this dependency and continues to seek vainly after solutions, this may be to the detrement of one's wellbeing. One compares himself with others and experiences his dependency as personal failure.[4]

This scenario proceeds from the assumption that the differentiation in the needs of the elderly will be clearly expressed.
This means that the elderly themselves must make known a very wide range of desires.
For the elderly who would like to continue to participate in the normal life of society, desires might include:
- more paid (part-time) jobs for the elderly;
- employing the elderly for teaching purposes;
- political participation as well as participation in the management of societies;
- volunteer work in the welfare sector, possible after having taken a course, for instance in infant crèches.[5]

On the other hand, many of the elderly will not be able or desire to participate in society to any considerable degree. There may be various reasons for this. In cases where poor health impairs

functioning, participation will of necessity be more limited.
Thanks to the fact that they are better informed and more assertive, the elderly will make higher demands, both as regards (health) care facilities and as regards accessibility to all sorts of general facilities which promote participation in social life (public transport, public buildings, mobile libraries, etc.).

4.2.2 Values and norms as applying to the elderly

It is assumed in the growth scenario that present values and norms with respect to the elderly will continue to be discriminatory: the disengagement approach with respect to the elderly by younger categories of the population (under the age of 55) will dominate.[6)]
The result of this will be that the wishes and desires of the elderly will be met not at all or only to a very limited degree. For instance, heavy pressure is placed on the elderly to leave the labour process prematurely, inter alia through early retirement. Up to the present large groups of the elderly in the age category 55-65 years of age have already quitted the labour process.[7)]
Moreover, the percentage of unemployed older people is fairly high. Assuming that there will be no significant improvement in the labour market situation till the year 2000, it would seem likely that in future too considerable numbers of people will quit the labour process before their 65th year. This will be intensified by the growth of the supply of labour (especially as a consequence of participation by married women in the labour process and the growth of the population in the past). Table 4.1 gives an overview of this.

It is assumed that no part-time jobs will be created specially for the elderly. Possibilities for retraining of the elderly - in as far as this training is provided in the context of the job - will also be very limited. The sum total of these tendencies is that the percentage of the elderly who participate in the labour process will be even lower than is now the case. The official pensioning age will continue to be 65.

Studies - both in the Netherlands and in other countries - show a negative relationship between unemployment and health. Studies organized in the Netherlands by the SCP as to the living conditions of the unemployed and disabled,[8)] show that unemployment can inter alia lead to psychic disturbances (such as heightening of fatalism,

alienation, misanthropy and dissatisfaction). Moreover, the unemployed experience their state of health less favourably than the employed. Furthermore, the unemployed show a greater tendency than the employed to heavier smoking and drinking.
Not only unemployment, but also the general discrepancy between the aspirations of the elderly and the manner in which these are met by society, leads to the heightening of risk behaviour.

Table 4.1 Trends in growth of the supply of labour, 1971-2000 (x 1000 work years)

period	particination in education	invalidity	married women	premature retirement	total	demography	migration	total labour supply
1971-1975	-220	-104	+ 77	- 1	-248	+257	+58	+ 67
1976-1980	-126	-128	+105	-82	-231	+334	+63	+166
1981-1985	- 33	- 42	+ 60	-17	- 32	+385	+26	+379
1986-1990	- 2	- 31	+ 65	-10	+ 22	+300	+15	+337
1991-1995	- 4	- 17	+ 48	- 7	+ 20	+114	+11	+145
1996-2000	- 4	- 17	+ 52	- 1	+ 30	+ 8	+10	+ 48

Source: SCP, 1984, p. 65.

4.2.3 Volunteer work

In this scenario, chances for the elderly to obtain paid work are thus limited. This would open possibilities for doing more unpaid work (volunteer work).
Various types of volunteer work can be distinguished:[9)]

- 'Traditional' forms for volunteer work, both active (for instance the Red Cross) and passive (e.g. organizations in the area of recreation); in the case of the passive traditional type of work, the great majority of participants assume a strictly consumerist attitude.
- Volunteer action, aimed at withdrawal from the abstract and large-scale society (e.g. self-help groups, small-scale movements).
- Volunteer action aimed at structural improvement of the living conditions, resulting in claims to government to introduce

measures for eliminating existing injustices. As opposed to the traditional forms of volunteer work, this type of volunteer action is often linked to political preferences.

In this scenario, the first and third types of volunteer work will be encountered most often among the elderly. The volunteer work performed by the elderly is also influenced by values and norms of society with respect to the elderly. This implies that no very large investments will be made to promote volunteer work (for instance the establishment and maintenance of a paid organizational framework), and that volunteer work may not exert a detrimental influence on employment goals by constituting a threat to professional workers. Forms of volunteer work which meet these criteria are:

- (management) functions in political parties;
- unpaid work for institutions with a religious or social function;
- unpaid work for interest groups (e.g. a union of the elderly);
- activities in societies and clubs devoted to hobbies, sport or conviviality;
- activities in the welfare sector (e.g. crèches, substitute grandparents for handicapped children in institutions).

A lot of the elderly will have difficulty accomodating themselves to these limited possibilities.

4.2.4 Decreasing solidarity

A decrease in solidarity, both between the generations and within the generations of elderly is one of the points of departure of the growth scenario. Solidarity refers particularly to the aid aspect.[10)]
In the case of intergenerational solidarity, in most cases this will be a one-sided relationship. Younger people will assist the elderly in case of illness or a decreased ability to manage alone. The people who provide aid may be (grand)children, other relatives, neighbours, friends or volunteers.
Four developments limit this form of solidarity:

(a) The increasingly individualistic lifestyle;[11)]
(b) Emancipation of women: participation in (professional) work decreases the possibilities for informal aid (umbrella care);
(c) Geographic mobility increases the distance between children and parents, as a consequence of which it becomes (much) more difficult to provide aid;

(d) By the time parents require assistance, the children too are often middle aged or even elderly, and may themselves have to cope with health problems.

Willingness to provide informal aid is moreover closely related to the type of assistance required. In cases where the elderly require (domestic) aid for a relatively long period, there is in general a consensus of opinion that professional aid should be called in.[12] In practice it is found that nearly 50% of the elderly under 80 years of age who can no longer manage alone receive informal aid, as against 30% of the elderly over the age of 80.[13] According to this scenario, these percentages will decrease.

The double ageing of the population, plus the objective of the elderly to continue living independently for as long as possible will mean that the need for (domestic) aid will increase, while relatively speaking, increasingly less of this aid will be provided informally. This means that there will be a sharp increase in the appeal made to professional aid (and possibly commercial aid). Moreover, the elderly often evince a preference for professional aid. They are then no longer dependent on the kindness of others. The aid they receive is then a right for which they pay, instead of a favour.[14]

Another implication of the above is that alternative residential forms, involving both the elderly and the young (kangaroo dwellings, group housing with a heterogeneous age structure) will not be forthcoming.

The individualistic lifestyle will also have consequences for intragenerational solidarity (solidarity among the elderly as a group). Older people will be inclined to exert their right to professional aid both for themselves and for their peers. An exception will be constituted by elderly people who form part of a household (usually partners). Since the traditional nuclear family is the main form of cohabitation encountered in this scenario, the number of (very) old people living alone will increase significantly (especially as a consequence of being widowed, see Appendix B) with a corresponding need for aid. Group housing with a homogeneous age structure are not expected to take hold to any significant degree.

In Chapter 6 the question of solidarity is gone into further as one of the disturbing developments. In this context the consequences of a dramatic decrease in solidarity, both qualitative and quantitative, are discussed.

4.2.5 Attitudes to sickness and health

A decreasing sense of responsibility for one's own health is one of the characteristics of the growth scenario. Various factors underlie this attitude:

(a) The discrepancy between aspirations of the elderly and the possibilities open to them will lead to increasing stress reactions. One of the consequences of this will be that there will be an increase in risk behaviour. To a greater degree than might be expected from the reference scenario (see Section 2.3), there will be an increase in the use of alcohol and tobacco. Moreover the way in which people cope with their problems may be expected to exert a significant influence on health.

(b) By comparison with the reference scenario, there will be a significant increase in tolerance of euthanasia in this scenario.
The most important reason for this is that medical and medical-technological developments create increasingly more possibilities for prolonging life, at the cost of the 'quality of life'.

(c) There will be a markedly heightened tendency to place responsibility for one's health at the physician's door; medical experts will enjoy a high degree of confidence, and patients will be prepared to follow their advice in as far as this concerns medication and treatment, but much less where it comes to dietary habits and all sorts of other aspects of 'lifestyle'.

(d) The psysician will also be confronted with health problems stemming from stress and dissatisfaction.[15)]

In this scenario only to a very limited degree will (government) measures help to promote more healthy lifestyles. These measures include (1) direct influencing of behaviour (for instance smoking prohibitions in public buildings), (2) indirect influencing of behaviour (placing obstacles in the way of risk behaviour, such as raising excise duty on alhocol and tobacco) and (3) health education and counselling.
Only to a very limited degree or not at all will the implications of these developments for the state of health of the elderly till the year 2000 be noticeable. The developments will come about gradually, and the time span till the year 2000 is too short for them to exert

any significant influence on the state of health of the elderly. It is expected that after 2000 their influence will be felt.

4.2.6 Increasing professionalization

The tendency to professionalization in health care and social services continues. In this scenario, the disadvantages of such professionalization do not outweigh the advantages.
The disadvantages quoted are: detached and businesslike aid, monopolization of knowledge and skills, and the self-interest of the professionals[16)] (cf. Section 5.3.1).
Aspects viewed as advantages are: independence of informal aid (with the often one-sided role of the elderly as recipients of aid) and qualitative improvement of (health) care.

Pressure to increase professionalization in (health) care derives from two sources: from the professionals themselves, and from the users of the professional facilities (including the elderly).

(a) Pressure from professionals
In Chapter 3, medical and medical-technological developments were described as these emerge from the literature and the opinions of the experts consulted by us. It was concluded that improvements will be especially in the sphere of palliatory measures (relieving the effects of disease without actually curing the disease) and less in the area of prevention and cure. 'Breakthroughs' of any significance for the most important diseases encountered among the elderly are hardly expected in the period under review. Only verly slow progress is expected per disease. New technological apparatus will improve possibilities for making diagnosis and implementing and monitoring therapy (scanning apparatus, lasers). In additon to this, the development of all sorts of aids is foreseen (prostheses, also for implantation, improvement of medication, including reservoir pills). See in this connection Sections 3.3.2 and 3.3.3.

The health care organization will exert considerable pressure to be given scope to apply the new possibilities in practice. This will almost always involve considerable costs. Whether a system of medical and medical-technological 'top care' will be achieved is a question of (government) policy. Chapter 7 goes into this 'supply' side of facilities.

A second tendency to increasing professionalization is evinced in the development of new specialisms. Two examples are here discussed in greater detail.
The first example came to the fore during the group consultations: the possibility and desirability[17)] of a new specialism 'pelvic medicine'. In this specialism the existing specialisms of urology, surgery, neurology and gynaecology would unite their knowledge and skills in order to make it possible better to deal with the ills of the elderly (incontinence, prolapse, etc.).

The second example relates to the specialism geriatrics. This specialism was recognized in 1981. Discussion surrounding its introduction was inter alia concerned with the question whether this specialism should be applied only in hospitals or also outside of hospitals. If geriatrics should obtain a place in general hospitals, should this be viewed as a function (without beds of its one) or as an independent department (so-called Geriatric Wards of General Hospitals - GAAZ)?
According to this scenario the experiments with GAAZs which are now carried out on a limited scale, will be expanded and eventually GAAZs will be introduced on a large scale.

Developments in nursing of the elderly constitute a third example of expanding professionalization (see Section 3.5.4). 'Nursing of the patient in bed' is replaced by 'nursing of the patient out of bed'. This could prevent a number of complications now encountered among the elderly, such as decubitus, contractions, urinary infections, incontinence, pneumonias and derangement. Nursing will develop from passive nursing to nursing focused on activation with the objective of maintaining the condition of the patient at the highest possible level. To this end, use will also be made of ergonomic principles. This development will have consequences both for nusing in intramural institutions and for district nursing.

(b) Pressure from the consumer
In as far as the emancipated, assertive, and well-informed elderly of the future suffer from health problems, they will desire to make use of the best facilities available.
This means that society will bring pressure to bear for the provision of 'top care'. People will fortify themselves (inter alia by the formation of unions) for the purpose of combating or doing away with restrictions governing treatment (for instance age limits for heart transplants). There will be a minimum of restrictions to the 'right to treatment'.

The consequence of all these developments will be that there will be a vast increase in the amount of treatment provided by the health care sector, with an accompanying increase in costs. This question will be discussed further in Section 4.4. A further problem is the friction which may arise, for instance between physician and patient, as a consequence of the discrepancy between the possibilities for making diagnosis and the possibilities for effecting cure. This problem is already encountered, but will increase in the future.

4.3 Consequences for facilities

The greater possibilities in the area of medicine and medical technology, as well as the high inclination to make use of health care facilities, and the reduction of informal aid, will, taken as a whole, result in a sharp increase in the demand for facilities. The increased demand will however not apply equally to all facilities. In the pages which follow this matter will be more closely examined. In so doing, a distinction will be made between health care facilities and facilities in the area of social services.

(a) Health care facilities

In the reference scenario, six health care facilities of which the elderly make use were discussed, namely: nursing homes and hospitals (intramural), day treatment (semimural) and GP care, ambulatory mental health care and district nursing (extramural).

How will the demand for these facilities develop on the basis of the points of departure discussed in this scenario?

Efforts of the elderly to continue to function independently for as long as possible will on the one hand be positively influenced by medical and medical-technological possibilities which may mainly serve to keep the elderly mobile for a longer period. To this end medical intervention will often be necessary. It is expected that the implication of this will be that the demand for treatment in hospitals will increase considerably, both absolutely and relatively (in other words, the increase will be greater than might be expected solely on the basis of demographic developments).

The creation of GAAZs could have the effect of reducing the use made of nursing homes. It is assumed that though there will be an absolute increase in the demand for treatment in nursing homes as a consequence of double ageing, the relative demand will remain constant. The nursing home will mainly acquire a long-stay function for patients requiring heavy nursing care for whom the chances of a cure are practically nil. Patients who may in any way profit from treatment will be admitted to the GAAZs.[18)]
In general, the emphasis will be on keeping the sojourn in intramural facilities as short as possible. This will have the effect of (greatly) increasing the demand for day treatment and district nursing, both absolutely and relatively.
The joint effect of the values and norms with respect to the elderly discussed in Section 4.2.2 which lead to increased risk behaviour and psychic disfunctioning, as well as the concepts with respect to sickness and health discussed in Section 4.2.5, will be an increased demand, both absolute and relative, for GP aid and ambulatory mental health care.

(b) Facilities in the social services sector
Three facilities are included under this sector: homes for the elderly (intramural), home help and adapted dwellings for the elderly (extramural).
It would seem fairly obvious that a facility such as homes for the elderly, which have been long disputed inter alia in connection with the fact that the total care provided by them leaves no scope for some degree of activity on the part of the elderly, will lose even more ground in this scenario. Moreover, even in the case of decreased mobility, residing independently is preferred. As a consequence of this, there will be a sharp increase in the demand for adapted dwellings for the elderly (inclusive of monitored dwellings). There will also be an increased demand for service flats. If this altered demand is met, this could result in a (sharp) decrease in the percentage of elderly persons who make use of intramural facilities.

Home help (both from traditional home helpers and from personnel who assist the elderly and are based in a home for the elderly) will have to increase considerably. This will mean that the district function of homes for the elderly will acquire an important position (cf. the eleven district-oriented functions as listed in Section 2.5.5). The demand for home help will increase even more as a result of the decrease of informal aid (from relatives, neighbours,[19)] acquaintances and volunteers). This could to some degree be compensated by making use of commercial domestic aid.

It is however questionable in how far the financial position of the elderly will permit of this. It is assumed that there will be a relative improvement thanks to better pensions, but nonetheless commercial aid on a large scale will not be within the reach of large groups of the elderly. At the moment, 16% of the elderly of 55 years of age and older (sometimes) have assistance from a paid domestic help, while 19% of people of 65 or older obtain such help.[20] Table 4.2 gives an overview of the above.

Table 4.2 Expectations with respect to the use of facilities by the elderly

	absolute use	relative use
GP care	strong +	strong +
amb.ment.health care	+	+
homes for the elderly	-	-
nursing homes	+	0
day treatment	strong +	strong +
hospitals	strong +	strong +
district nursing	strong +	strong +
home help	strong +	strong +
adapted dwellings[21]	+	+

See notes 5 and 6 of Chapter 1 for an explanation of the signs and the concept 'relative use'.

4.4 The economic context

The central question in this section is how the assumptions with respect to economic growth or shrinkage will bear up to the scenario presented here.

Economic growth is per definition essential for this scenario since for nearly all facilities (with the exception of homes for the elderly and nursing homes) there will be an absolute increase in demand which exceeds the increase which would result from (double) ageing alone.
In the reference scenario (see Section 2.6) it was established that in the period 1981-2000 a growth of the national income of 0.6% per annum would be necessary to meet the increase in demand arising from

demographic developments alone.[22)] The growth of the national income in the growth scenario should be considerably greater. A quantitative indication of what this growth should be is however not possible.

It may be expected that - even in a situation of (considerable) economic growth - the discussion as to controlling costs in health care will continue. At the moment, collective expenditure for health care accounts for approximately 10% of the national income. Generally speaking, this is viewed as a maximum. The increase of costs in the past, partly as a result of the enormous growth in the number of treatments, has been explosive (see Tables 4.3 and 4.4) and in future government will put a stop to this growth (cf. the various norms for health care facilities).

Table 4.3 Total expenditure for health care

	1963	1968	1973	1978	1980	estimate 1983
milliard guilders	2.3	5.3	12.3	23.5	27.8	35.3
% gross national product	4.5	5.9	7.4	8.4	8.8	
% intramural approx.	40	44	54.5	58.7	58.6	

Source: Van der Meer, 1983, p. 4.

Table 4.4 Estimated number of treatments performed in hospitals and outpatient departments in the period 1968-1978

	1968	1978
number of operations	600,000	1,600,000
number of deliveries	67,000	91,000
number of x-ray diagnoses	3,200,000	6,200,000
number of functional tests	1,200,000	2,500,000
number of physiotherapy treatments	5,300,000	7,000,000
number of laboratory tests	69,000,000	270,000,000

Source: Van der Meer, 1983, p. 6.

In how far the attempts to control costs will lead to problematic situations (for instance waiting lists as a consequence of capacity restrictions) is difficult to judge.

In a situation of economic shrinkage this scenario will result in serious problems. The pressure to expansion of facilities will be considerable, both from professionals and from society. The problems which will arise will be not only of a financial nature, but also relate to various other aspects of the question of distribution: who has access to the (health) care system, where do priorities lie with respect to maintaining facilities, etc.

The reader is referred to Table 1.1 in Chapter 1 for a comprehensive overview of the growth scenario.

Notes Chapter 4

(1) This in spite of attempts by government precisely to reduce the demand made on all sorts of facilities (especially intramural facilities). As was already indicated in the introduction to this report, in scenarios B and C no account is taken of government policy and its influence on the supply of facilities.

(2) There is no unambiguous definition of dependence versus independence. See Van den Heuvel 1976, pp. 162-172 for a discussion of this point.

(3) Scientific Council for Goverment Policy (WRR), 1982, p. 177.

(4) Kronjee, 1984, p. 83.

(5) United Nations, 1980. Quoted in WRR, 1982, p. 179.

(6) It is assumed that till the year 2000 the 'emancipation struggle' by the elderly will continue. This struggle will eventually result in the abrogation of all sorts of discriminatory laws and measures (for instance the fixed retirement age of 65).

(7) In 1981 there were still 322,000 employed persons in the age category 55-65 years of age (exclusive of independently employed persons). Minister of Welfare, Health and Culture et al., 1983, p. 6. Premature retirement is generally approved from the age of 62 of 63.

(8) Quoted in SCP, 1984, p. 73.

(9) Adriaansens and Zijderveld, 1981, pp. 73-75.

(10) A large number of analyses have appeared on the relationship welfare state - individual. It is repeatedly pointed out that the far-reaching 'care arrangements' have reduced the solidarity between citizens and have led to a consumerist attitude, aimed at pursuing the personal identity of the individual (qualitative individualism). See for instance Schnabel, 1983, pp. 25-67. Other authors speak of an immoralistic ethos which dominates in the welfare state (Adriaansens and Zijderveld, 1981).

(11) A recent study as to the residential desires of the elderly carried out by the Housing Research Institute in Delft showed that in general young people have no desire for contact with the elderly. People are completely occupied with their own activities. See Houben, Wind and Moeskops, 1984.

(12) SCP, 1984, p. 377. In a survey carried out in 1983, 63% of interviewees were in favour of professional aid in such cases, while 37% were in favour of informal aid.

(13) De Groot et al., 1981.

(14) As opposed to this - according to the COSBO - the interests of the elderly do not always run parallel with those of professional providers of aid.

(15) This represents the tendency to attach the label 'sick' to increasingly more phenomena, which implies that the persons concerned become absorbed in the circuit of health care facilities.

(16) A further negative aspect of health care is iatrogenesis: the appearance of new illnesses brought about by physicians, hospitals and drugs - the corner-stones of the medical system. In other areas of wellfare, such as social services, the same phenomenon is called therapeugenesis. Achterhuis, 1980.

(17) We will not enter into discussion on the question whether and in how far this is an example of professionals furthering their own interests.

(18) Criteria for determining whether possibilities for 'cure' still exist will be difficult to pin-point.

(19) Possibilities for help from neighbours will only exist if they themselves do not suffer from ADL problems.

(20) CBS, 1984, p. 106. In 1976 the percentage was 11 for people of 55 and older and 13 for people of 65 and older. CBS, 1983, p. 62.

(21) See Section 2.5.11 on the housing situation (inter alia in connection with terminology relating to various types of dwellings for the elderly).

(22) Medium variant of the population forecast. It is assumed that the ratio between the national income and collective expenditure will remain constant.

5 Scenario C: Decreasing growth in demand for facilities

5.1 Introduction

In this scenario a picture is sketched of various more or less consistent developments which, provided there is an adequate interplay with supply, could lead to a decrease in the (growth of) demand for facilities. The following developments are discussed in this chapter as being the most important which could have this result.

(1) Altered attitudes to sickness and health (greater responsibility for one's own health resulting in a more healthy lifestyle).
(2) An altered socio-economic position of the elderly (work, pensioning, income).
(3) Changes with respect to the concept of care (de-professionalization, de-instituationalization, etc.).

Items (1) and (2) are dealt with in Section 5.2 (Social developments). As far as the social trends which are not mentioned here are concerned (emancipation of the elderly, educational level, attitudes with respect to (active and passive) euthanasia, political significance of the elderly, etc.), it is assumed that they will continue to develop in the same direction as described in the reference scenario (Section 2.4).

With respect to the state of health of the elderly and medical-technological developments too, in this scenario no shifts are expected which differ to any significant degree from the situation described in Section 2.3 and Chapter 3. Though the attitudes described in this scenario with respect to sickness and health, lifestyles and socio-economic position may be regarded as important determinants of the state of health, their influence on the state of health of the elderly till the year 2000 is viewed as marginal. The reason for this is that even though significant changes should come about in these determinants, the influence they will exert on the elderly will be less than if this group had been exposed to these changes from a younger age.
Viewed over a longer period of time however, the consequences of these shifts for the health of the elderly will become more visible. This is briefly discussed in Section 5.2.4.

The most important trend which determines the decreasing pressure on facilities in this scenario is consequently not so much the (improved) state of health of the elderly, but rather changes in the concept of care. Section 5.3 discusses this topic at length. In Section 5.4 the relationship between this scenario and (alternative) economic context(s) is outlined.

The reader is referred to Table 1.1 in Chapter 1 for an overview of this scenario.

5.2 Social developments as they affect the health status

5.2.1 Altered attitudes with respect to sickness and health

In this scenario it is assumed that the importance which people attach to health - after a slight decrease in the eighties (see Table 5.1) - will increase further at the cost of, for instance, a good income.
This shift goes hand in hand with a shift in general social values and norms: from a society oriented toward achievement, work, productivity and an attachment to material goods, to a society where immaterial values (health, relationships, self-improvement etc.) play an increasingly important role.

In addition to the fact that increasingly priority is given to health, people are becoming increasingly of the opinion that health is to a large degree influenced by one's own behaviour. Consequently, this scenario proceeds from the assumption that people will increasingly accept responsibility for their own health. This means that the 'engineering era' and the 'medical era' are followed by the 'post-medical era'.[1] In the 'engineering era', inducing the physical environment to meet the minimum standards for health (sufficient food, drinking water, hygiene), constituted the key to progress. This era was followed by the 'medical era', in which allopathic medicine constituted the dominating approach to health, based on mass vaccination and extensive use of antibiotics. In the 'post-medical era', it is realized that the place of (allopathic) medicine in relation to the most important determinants of health, such as individual behaviour (smoking, lack of exercise), social organization (stress), economic status (poverty, over-consumption) and physical environment (pollution), is only limited.

Table 5.1 Things which people consider important in life 1966-1983 (in percentages)

	1966	1975	1979	1980	1981	1983
good income	3.3	4.1	5.7	8.7	7.5	2.6
happy family	7.7	13.4	11.9	13.5	13.4	14.8
health	35.4	43.5	50.8	42.6	48.0	51.9
much free time	0.5	0.3	0.5	0.4	0.2	0.9
happy marriage	34.6	27.4	18.7	18.0	17.4	20.8
religion	15.4	8.6	5.0	5.2	5.7	4.5
satisfying work	2.4	1.4	3.8	4.4	2.7	1.7
many friends and acquaintances	0.9	1.4	3.6	7.1	5.2	2.8
N	1.744	1.744	1.865	1.853	1.832	1.877

Source: Survey of Religion in the Netherlands, 1966.
Cultural Changes is the Netherlands, 1958-1975, 1979, 1980, 1981, 1983. Quoted in: SCP, 1984.

In this scenario high priority is accorded to primary prevention, especially by alteration of lifestyles. Alteration of lifestyles is further motivated by intensification of measures such as (1) direct influencing of behaviour (e.g. by smoking prohibitions in public buildings), (2) indirect influencing of behaviour (creation of obstacles to risky habits, for instance by raising excise duty on alcohol and tobacco), and (3) education and counselling (inter alia medical advice to the elderly). These changes in behaviour and attitudes to health will emanate primarily from the middle classes. In the past too the middle classes were often the pioneers of general social developments which led to improvements in health.[2)]

It will have become evident from the list of determinants given above, that there are limits to the degree to which the individual can be responsible to his own health. This responsibility relates only to 'self-imposed risks'. Limitation of personal risk behaviour should be accompanied by changes in other areas which affect health, such as sharpening of requirements with respect to conditions of work, environmental pollution, food production, etc. Epidemiological research and health monitoring could function as important aids in this respect. Thanks inter alia to increasing possibilities in computer technology it would be possible to determine at an earlier stage deviations in patterns of health and to detect the possible causes.

Ex ante evaluations in the area of health will also gain in importance, just as at the moment considerable attention is devoted to technology assessment (TA) and environmental impact assessment (EIA). In making a health impact assessment, important public and private plans will be closely examined as to the positive and negative effects which they may be expected to exert on health.

As was already stated in the introduction, it is however expected that only after the year 2000 these developments will lead to significant changes in the state of health of the elderly. This subject is briefly discussed in Section 5.2.4.

5.2.2 Work

In this scenario, the most salient problem of the eighties, unemployment, is increasingly viewed as a problem of distribution rather than of economic stagnation. If unemployment is approached from the latter angle, measures will be aimed at raising economic activity to a higher level (e.g. by means of wage restraints to make it possible for the yield of companies to revive, and/or by means of reduction of taxes and/or technological innovation).
In the nineties, this option is increasingly deserted in favour of rigorous shortening of working hours, in accordance with the view of unemployment as a problem of distribution. In the future this choice becomes increasingly urgent in view of the growing supply of labour. Calculations of the Central Planning Bureau show that the labour supply in the eighties will be triple that in the seventies, namely 716,000 work years as against 233,000 work years. In the nineties a further increase of 193,000 work years is to be expected.[3)]

The increasing supply of labour can be explained by the increasing participation by married women in the labour market, and also by purely demographic developments: for instance in the eighties there is a net increase of 685,000 persons on the labour market.[4)] In this scenario it may be expected that the elderly too (though initially only to a limited degree) will participate more in the labour market. The more active and emancipated elderly will view the set age criterion for withdrawal from the labour process as discriminatory and stereotype since a person's calender age and functional age are not necessarily synchronous. This would fall in line with developments which have been operative in other countries for a much longer period, for instance the Age Discrimination in

Employment Act in the U.S., which forbids dismissal on grounds of age. At the moment this act is not (yet) completely prohibitive (dismissal on grounds of age after the age of 70 is permitted), but this limit too is under discussion.

In the period till 2000 the number of working hours per week will gradually drop to approximately 25. This implies a much greater reduction in working hours than would be in accordance with past trends (between 1960 and 1983 the average number of working hours per week dropped from 45 to 40). The form in which reduction of working hours takes place varies per branch of employment (reduction per day, per week, per year, etc.), but in most instances the five-hour work day will be introduced. This offers possibilities for a better division of domestic tasks and care of children and the elderly over the sexes. Furthermore, it opens possibilities for the elderly to continue working longer since a five-hour work day is less tiring than an eight-hour day.
Moreover, the elderly who have (for some time) before pensioning operated with the five-hour work day, will be less worn down (this applies especially to physically strenuous jobs), and will consequently be less inclined than is at present the case to retire prematurely from the labour process. Also, computerization offers less mobile elderly people more possibilities for continuing to participate in the labour process, since it is expected that working with computers will become increasingly decentralized. More and more, this type of work will take place at home or in small-scale decentralized offices via electronic networks.

Since paid work will assume a less prominent position in society, and there will be more time for unofficial work, a process will come about whereby paid and unpaid work will be evaluated increasingly on the same footing. This could exert a profound influence on the outlook with regard to the elderly, since it is often argued that the problems of the elderly in our present-day society have their roots in the far-reaching economic concept of labour: the elderly are primarily an economic category, the inactive, the unproductive. The prospect sketched above largely eliminates the grounds for this view of the elderly.

In addition to the shortening of the work week, possibilities will be created on a larger scale than is at present the case for 'sabbatical leave'. For our purpose especially the following forms of leave are relevant:

- educational leave for further schooling or re-schooling;

- leave for the purpose of caring for ill or dying partners or parents;
- leave preparatory to (flexible) pensioning.

The consequences for incomes of the shortening of working hours will to a large extent depend on the economic context in which this takes place. In times of economic growth it will weigh less heavily on individuals than in times of economic stagnation. In the most unfavourable situation it is assumed that reduction in working hours will result in an equivalent reduction of income. In recent years, willingness to accept less income has increased among proponents of shorter working hours.[5)]

A number of factors which are linked to shortening working hours could however exert a positive influence on the development of incomes, namely:

- Better use of the supply of capital. This entails the assumption that a reduction in working hours will be coupled with an extension of hours of operation (for instance two work-shifts of five hours each per day).
- Decrease in umemployment and consequently of unemployment premiums.
- Increase in the number of people employed, which will mean that families or households will often have more than one (reduced) salary, so that there will be less of a decrease or even an increase in the total family or household income. Whether this will already apply to the next generation of elderly is however doubtful, since the family structure has usually been based on one (male) breadwinner. As a consequence, a financial gap may come about between the coming generations of elderly: traditional households with one breadwinner versus households based on two breadwinners.

A reduction of working hours might however be linked with a number of developments which would cause a less rapid decrease in unemployment than is expected from this measure (for instance the shorting of working hours encourages more people to enter the labour market, or people take more than one job). When it is posed that the shortening of working hours encourages more people to seek jobs, it is important which definition of unemployment is adhered to. If people who are not registered as looking for work are counted as unemployed, the possible reduction of the number of unemployed will probably be greater than if only the officially registered are counted.

5.2.3 Pensioning

As was already suggested above, in the period till 2000 possibilities will be created for flexible pensioning. This means that the compulsory pension age of 65 will fall into abeyance, and that the moment of retirement will be determined by the employee himself between the ages of 60 and 70. The official premature pensioning regulations will also be discarded. With this, developments in the Netherlands would fall in line with developments which have already made their appearance in other countries (U.S., Federal Republic of Germany, France, Scandinavia, United Kingdom).

In addition to flexibility in the age at which one retires, more possibilities are created for 'half pensioning' and 'gradual pensioning', the (half) pensions being supplemented by incomes from part-time work.

In the first instance, expansion of possibilities for flexible pensioning (and especially the aspect of continuing to work longer) would appear to be at odds with the trend at present encountered in society to retire prematurely from the labour process (that is to say before the retirement age of 65), partly by voluntary premature retirement, and partly through unemployment or being declared medically unfit.
At the moment, premature retirement is strongly influenced by social and economic motives: to make way for young unemployed persons who often receive lower wages and whose productivity is sometimes greater than older workers.
Apart from the effectivity of this for combating unemployment in the short term, its effectiveness in the long term is questioned. Considerations which may lead to a reversal of the short-term policy of the eighties and the introduction of flexible pensioning, are:

(1) **Economic-demographic considerations**: the increasing ageing of the population signifies a greater share of economically non-active persons in the population. The dependency of the economically non-active on the economically active is usually expressed as the 'demographic burden':

$$\frac{\text{0-19 year olds + 65-plussers}}{\text{20-64 year olds}} \times 100\%$$

Actually, to show more correctly the implications of no reversal in the trend to early retirement, it would be more realistic to substitute the age of 55 for the age of 65 in the formula. The following picture would then be obtained till the year 2030 (see Table 5.2).

Table 5.2 Demographic burden till 2030 (medium variant)

	1980	1990	2000	2010	2020	2030
demographic burden	109%	93%	95%	107%	118%	132%

Source: Calculated on the basis of CBS, 1982.

The demographic burden begins to increase from 1990 till by 2010 it has attained the level of 1980, and after 2010 surpasses the level of 1980. Even though the burden in 2010 is the same as in 1980, the composition of the numerator is completely different. The young in the numerator are increasingly replaced by the elderly. This has important financial implications, since the costs of the elderly are estimated at 2.5 to 3 times as high as for the young.[6] The 'burden of the elderly':

$$\frac{\text{55-plussers}}{\text{20-54 year olds}} \times 100\%$$

would perhaps give a truer picture (see Table 5.3).

Table 5.3 Burden of the elderly till 2030 (medium variant)

	1980	1990	2000	2010	2020	2030
burden of the elderly	43%	42%	46%	58%	70%	81%

Source: Calculated on the basis of CBS, 1982.

From a point of view of financial considerations (payability of benefits) and a shortage[7] of labour supply, flexible pensioning might become important. The degree to which flexible pensioning can meet the demographic developments described above is naturally heavily dependent on the willingness shown by the

elderly to continue work. Experience with flexible pensioning in other countries is not very favourable.
Calculations[8] have shown that flexible pensioning, viewed on the macro level, will not lead to any substantial increase in costs, even though everyone should quit the labour process at the age of 60 to 65. Depending on the calculation model used, extra costs will vary between 0.5% and 3.5% of the total of salaries of the active population.
An experiment in the Netherlands with the Postal and Telecommunications Service also showed that there is little interest for continuing work till the age of 67.
Actually, it is not unlikely that the altered situation brought about by a five-hour work day and the financial necessity or attractiveness of continuing work will bring about considerable changes in attitudes.

(2) **Ethical considerations:** dismissal on grounds of age should be considered discriminatory; the elderly have as much right to work as every other person; the unemployment problem of the youth should not be one-sidedly shifted to the elderly (see also ILO recommendation No. 162). As a consequence of the increasing influence of inter alia unions of the elderly, these ethical principles will increasingly more be considered self-evident.

Concluding, it may be posed that though in the short term, individual desires (for instance continuing longer in the labour process) are at odds with social requirements (for instance, making way for young unemployed persons), nonetheless over the longer term it is not inconceivable that these considerations will increasingly complement each other.

The question remains in how far possibilities will exist for the elderly to continue to work longer in view of the opinion that they are less productive. (American) studies have shown that the assumption of a relationship between age and productivity is mainly mythical. For most jobs and occupations, the knowledge and skills of the elderly would continue to suffice till past the age of eighty.[9]
The wear and tear of work will also become less in the future as a consequence of shortening of working hours. The 'relative ageing' - that is, 'ageing' resulting from technical-economic developments - will be also reduced thanks to the earlier-mentioned 'sabbatical leaves' with which work will be regularly alternated.

Lastly, there is the possibility of the creation of so-called senior jobs. In Canada and the U.S. experiments with these jobs are already carried out on a large scale. After undergoing various tests, it is established what job best suits the 'profile' of the older person in question.[10)]

As regards the no longer employed elderly, a distinction should be made between general old age pensions and retirement pensions. It is expected that in the future, more people will receive a retirement pension (see also Section 2.4.7). Whether the level of these pensions will also increase depends on the question in how far reduction of working hours results in a substantial lessening of (the last-earned) wage, and whether the family was formerly dependent on one or two wage earners. In addition to this, the type of pension regulation plays a role (for instance in the case of reduction of working hours, the last-earned wage regulation is less favourable than other regulations).

Uncertainties also surround general old age pensions. Recently there has been considerable speculation in publications on the question of future old age pensions. The disadvantage of such publications is that calculations are often carried out on the assumption that other conditions will remain the same as in the base year. The advantage of scenarios is that other conditions are assumed also to change. An attempt is made to sketch a consistent picture of the changes which thus come about. This scenario shows the income position of the elderly as being less uniform than in the eighties:[11)]

(1) income on the basis of a full-time job (25-hour work week) (55-70 year olds);
(2) income on the basis of (flexible) general old age pension (60 year olds and older);
(3) income on the basis of (flexible) general old age pension and retirement pension (60 year olds and older);
(4) income on the basis of (halved) retirement pension plus old age pension plus part-time job (less than 25 hours per week) ('half pensions') (60-70 year olds);
(5) (compulsory) pension and/or general old age pension (after the age of 70);
(6) income on the basis of other than general old age pension (disability benefit, unemployment benefit etc.) (55-60 year olds).

In this scenario we proceed from the assumption that the income position of the majority of the elderly in the future will improve by comparison with the past, certainly for households with more than one wage earner. Only for the categories 2 and 6 is it - centainly in times of economic recession - not unlikely that the income position will deteriorate somewhat. Significant reversals in pension are however not considered likely in view of the increasing political engagement of the (better educated) elderly.

The income position of the elderly is of importance for the degree to which they make use of facilities and the type of facilities they use. This question will be discussed further in Section 5.3.

5.2.4 Long-term consequences for health

The consequences of these developments for the state of health of the elderly will probably only be visible after a long period of time. Generally speaking, the trends described could result in an improved state of health for the elderly, inter alia as a consequence of:

(1) a reduction of risk behaviour in combination with collective measures for promoting health;
(2) a reduced work load (25-hour work week) and better division of tasks over the sexes;
(3) better adjustment to individual desires (inter alia by means of flexible pensioning), possibly resulting in a heightening of the mental and physical wellbeing of the elderly;
(4) an improved income position. As various (American) studies have shown[12], the income position correlates positively with the 'active life expectancy', that is to say the number of years that the elderly may be expected to be able to perform ADL without assistance. In the age category 65 to 69 year olds, the differences in years of activity gained by improved income varied from 2 to 4, while for people of over 75 years of age one year gain was found. Incidentally, with studies of this nature, it might be questioned in how far the relationship between income and health should be attributed to variables such as lifestyle and conditions of work which correlate with income.
(5) a possible decrease in unemployment. If the introduction of reduction of working hours exerts a positive effect on unemployment in this scenario, this could have important consequences for health, since studies show a negative correlation between unemployment and health. The reader is referred to Section 4.2.2 on this matter.

There would be an improvement in the state of health of the elderly especially if diseases with a relatively long duration and which make no significant contribution to mortality (such as chronic rheumatism of the joints) could be avoided or postponed by the developments discussed above. Sickness would then be compressed into a shorter period before death. In the case of prevention and postponement of the diseases mainly responsible for death, which are mainly characterized by a relatively short (clinically manifested) period of illness, paradoxical effects could result for the health of the population, since if death is postponed, in the years gained there is a relatively greater chance of disabling diseases becoming manifest.[13] It is consequently difficult to give one overall picture of the state of health of the elderly in the long term as a consequence of the developments discussed in this scenario.
As regards life expectancy of men and women in the long term, in this scenario it is assumed that as a consequence of the increasing similarity in patterns of behaviour between men and women,[14] there will be a greater decrease in the gap in life expectancy existing between the sexes than in the reference scenario (see Section 2.3). This has important long-term consequences for the umbrella care with which partners provide each other (see Section 5.3.2).

5.3 Developments relating to facilities for the elderly

5.3.1 Central points of departure

This scenario is characterized by (strongly) changing concepts with respect to the manner in which care of the elderly is organized. In this first section the general points of departure of this changing concept of care will be outlined. The altered vision of care is characterized by:

(1) **Opposition to the continuing professionalization of care.**
Financial reasons are often quoted for this opposition: care will consume an increasingly large share of public funds, partly due to the fact that the care sector is less suited to rationalization and automation than other sectors. Care is primarily 'devoting time to someone'.[15]
In this scenario however, finance is not the dominating motive, though it is not absent. An intrincsic value is accorded to counteracting professionalization of care, since if the pressure to reduce professionalization does not come from inside, there

is considerable likelihood that limitation of professional aid will cause an increase rather than a decrease in demand. In that case one is inclined to think: 'Always ask for professional aid: you never know if you will get it.'[16)]
Proponents of reducing professional aid point to several disadvantages of professionalization. Professional help is characterized as detached and businesslike, as monopolizing certain knowledge and skills, and not being completely free of self-interest. Though in this scenario professional care certainly will not disappear, there will be a decrease in the ratio of professional aid to other forms of care (self help, umbrella care and volunteer work).
Personal responsibility for caring for one another, certainly in cases where aid can be (equally well) provided by non-professionals, will increase. Both sexes will participate in providing aid. As a result of the shortening of working hours the (free) time necessary for this will come available (Section 5.2.2). Thus, in this scenario 'time to care'[17)] has a double meaning: 'it is (again) time to care for each other', and 'there will be time enough available for looking after each other'.

(2) Closely related to the above are the efforts being made to **restrict admission of the elderly to the intramural sector.**
At the moment, approximately 11% of elderly persons over the age of 65 are institutionalized (8.1% in homes for the elderly, 2.8% in nursing homes). This percentage however gives an incorrect impression of the extent of this phenomenon, since it is only a momentary picture of the situation of the group of elderly. It gives no insight into the chances for an individual to be admitted to a home for the elderly or a nursing home in his lifetime. This chance is much higher than the 11% given. Calculations[18)] show that approximately 35% of all elderly persons in the Netherlands will eventually be admitted to a home for the elderly or a nursing home. Despite the large difference between homes for the elderly and nursing homes as regards numbers admitted at one moment, little difference is found to exist in the chances of ever being admitted to one or other of these institutions.
In the context of de-institutionalization, efforts are made to enable the elderly person to remian in his or her own residential environment for as long as possible. This stimulates the elderly person to activity, which would serve to promote both physical and mental wellbeing (cf. social component of dementia). Negative effects of hospitalization (loss of privacy and hospital syndrome) are also kept to a minimum.
The tendency to de-institutionalization is coupled with efforts to make it possible for older people to die at home. At present,

it is estimated that only 10% of the elderly die at home.[19] The possibility for a partner and/or children to get leave to care for the patient (Section 5.2.2) would exert a positive influence on this situation.

Institutions will however not disappear altogether. This applies especially to institutions which perform highly qualified and specialized jobs. In such instances there will however be a tendency to reduction of scale and expansion of the range of possibilities for semimural care.

(3) **Diversification of care.** In the past the care of the elderly - especially in the intramural sector - was often offered as a 'total package', both with a view to efficient functioning and on account of a certain reluctance on the part of the elderly to perform tasks which would still have been within their powers. The higher the amount paid by the patient, the greater the reluctance to lend a hand. In this scenario in the future care will be more 'cut to measure', and choice possibilities will be enlarged. The increase of 'unconditional income transfers' (Section 5.4) could be of considerable importance in this connection.

(4) **Repression of the tendency to label all problems as medical.** With increasing secularization, as a consequence of which the role of the church in dealing with social problems declined, and as a result of the (too) high expectations of medicine entertained by the public, the physician (GP) is increasingly confronted with these problems.
In this scenario an effort is made to suppress this increasing tendency. The limits of medical treatment will be confined more to questions involving medical knowledge (scientific purity in medical practice).[20] Social problems will be relegated to other (already existing) professional and non-professional facilities.

(5) **Upgrading of 'care'.** Trained in a system in which the emphasis is on the curing of disease, the status of health care workers, who devote themselves to the often lengthy and unsuccessful treatment or nursing of chronic diseases (care), is lower than that of those who apply themselves to treatment and nursing of acute diseases (cure). (Double) ageing will mean an increase in the number of chronically ill patients in the population. In order to be able to meet the demand for care which this will involve, it is not inconceivable that efforts will be made to upgrade occupations in this sector (for instance by recognition of a specific training).

The altered vision of care of the elderly sketched above will have consequences for all facilities, though some will be affected more than others. The most radical changes will probably take place in areas where there is most difference of opinion regarding the demand for aid and need for care and the solutions for meeting this demand.[21] The range of difference of opinion is shown in Figure V.1.

Figure V.1 Range of differences of opinion regarding demand and need for care and the solutions for meeting this demand

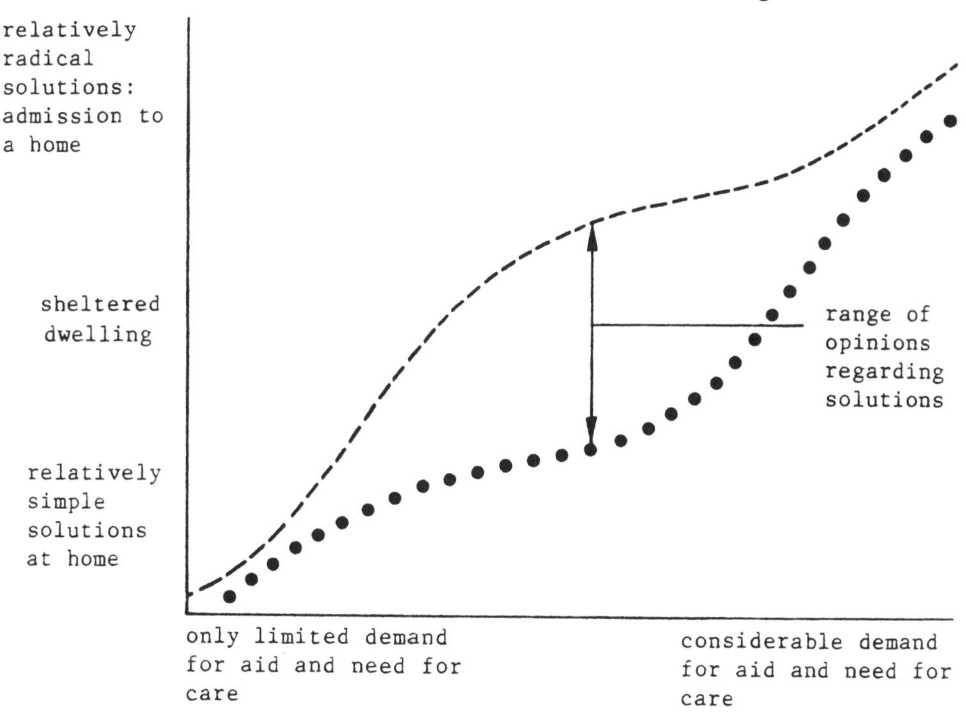

Legend:

-----: outlook of those who are primarily inclined to choose a radical solution, namely admission to a home

.....: outlook of those who are primarily inclined to choose a simple solution and try to make it possible for the elderly person to remain at home.

Source: Houben, 1984.

The horizontal axis shows the degree of the demand for care and the need for care. The vertical axis shows the solutions which may be chosen, ranked according to how radical are the solutions. A more or less direct relationship would be expected between the two variables. This is however not evident. Where there is the greatest divergence, in this scenario, as s consequence of a different attitude to care, changes will be greatest by comparison with the present situation.

In Sections 5.3.2 and 5.3.3 changes with respect to facilities wil be discussed. For this purpose facilities are classified according to their (most obvious) funtions: (1) sojourn or residential function, (2) care function (both care of the individual and domestic aid), (3) nursing function, and (4) medical treatment (diagnosis and therapy). The changes in the first two functions are greatest by comparison with the present situation. This is partly to be ascribed to the less specialistic nature of the care. On this account these two funtions are accorded relatively most attention in this scenario (Section 5.3.2).

5.3.2 Sojourn and care function

The most important facility for the elderly in our present-day society where the sojourn and care functions are combined is the home for the elderly. In the context of this scenario, this home in its present form will have almost disappeared in the future. As a result of the increasingly strict system of indication for admission, and the growing need for co-operation, homes for the elderly and nursing homes will become increasingly similar. In the long run this will have the result that there will be one intramural long-stay facility for elderly persons who are heavily dependent on assistance. This facility will offer diversified 'packages of care' suited to the specific needs of the residents. This however does not necessarily imply a trend to expansion of scale and centralization.

This long stay facility would at the same time serve as a headquarter for providing assistance to elderly people in their own homes (meals, socio-cultural activities, alarm systems, care during the night, day care and treatment, admission in vacation, (therapeutic) guidance of non-professional aid, etc.). The range of care and residential forms for elderly people who require less help will increase. These will include:

(1) The traditional family with marriage partner and/or children. The closed family will however increasingly make way for the open family. The monopoly position of family life will be increasingly undermined.[22)]

(2) Enforced living alone as a result of the loss of a partner or children leaving home will become less common thanks to the above. Long-term, in this scenario the decreasing gap in life expectancy between men and women too will lead to a reduction in the number of people living alone. This will have interesting consequences for the demand for facilities: in principle partners can care for each other longer, and loneliness, an important factor in the desire to be admitted to a home for the elderly, will be considerably reduced.
Nonetheless, the phenomenon of living alone will not disappear.

(3) LAT relationships and other forms of unmarried cohabitation of partners will increase as a consequence of the growing tolerance of these alternative forms of living together.[23)]

(4) In addition to alternative partner relationships, there will be the advent of larger primary groups. One example might be a self-chosen 'friendship network',[24)] in which members of the family too might participate. When such groups decide to live together in a commonly shared residential unit (however maintaining a large degree of privacy), we might speak of communes. Otherwise than in homes for the elderly, in these communes fellow residents have been consciously chosen. This factor becomes increasingly important as the individual becomes less mobile and must more and more fall back on the assistance of fellow residents and people in the neighbourhood.
Communes may have a relatively homogeneous age structure (senior communes) or be heterogeneous in age structure. In both instances the residents provide each other with aid. The size of the groups often makes jobs lighter than in the traditional nuclear family.[24)]

(5) Sheltered dwellings. This dwelling form differs from the other forms discussed in that professional aid is available. The professional aid is usually provided by an institution: specialized aid can be called in from the institution. In the future these residential forms would have outgrown the present experimental stage.[25)]

In all these residential and care types (though to a lesser degree in the case of (5), aid can be supplemented from outside if the (reciprocal) aid becomes too difficult as a result of lack of skills, or if the tasks involved are too heavy. The traditional forms of professional extramural aid for the elderly are home help and district nursing. This formal aid may be supplemented by informal extramural aid from children who no longer live at home, neighbours and volunteers.

In this scenario, no further increase is expected of the informal help provided by children living either at home or elsewhere. Reciprocal aid (the first point of departure) will be more a question of intra than intergenerational solidarity. Though, viewed objectively, the amount of time available for caring for the elderly will increase as a result of the shortening of working hours, many tendencies point in the direction of a decrease in help from children, such as increasing geographic mobility, reduced marital fertility, etc.
The elderly too find the asymmetric help relationship with children increasingly unsatisfying. The old three-generation family will not reappear. Moreover, research has shown that the nostalgic references to this type of family are mostly based on fiction. This form of cohabitation was often enforced by economic factors. The child who had shouldered the care of an elderly parent was repaid by being bequeathed the possessions of the parent.[26]

The prospects for (organized) volunteer work are viewed more favourably in this scenario. This refers to volunteer work both by younger people and by the elderly themselves. The following factors are expected to play a role:

(1) The Living Conditions Survey 1982[27] shows that 13% of older men (> 55 years) and 9% of older women would be pleased to perform (more) volunteer work. These figures relate to people residing independently. The figures for people in homes are 3 and 4% respectively. In how far this answer is given because it is socially approved, and what might be the obstacles to performing this work, is however unclear.

(2) The degree to which volunteer work is performed is heavily dependent on the level of education and the sector in which one is employed. Highly educated persons and persons employed in the quaternary sector are strongly represented in volunteer work. With a rising level of education and an increasing share of the quaternary sector in the national economy, an increase in volunteer work would also be expected.[28] In the light of present policy it would however not appear as though there will be an expansion of the quaternary sector in the near future.[29]

(3) Increasing shortening of working hours in principle creates more free time for performing volunteer work. Paradoxically enough, however, the very peoply who have no professional occupation, and of whom it might be expected that they would thus have a considerable amount of free time, are underrepresented in volunteer work.
In the case of the unemployed and people unfit for work, this might be connected with obstacles to volunteer work relating to the conditions for receiving unemployment grants.[30] Van Luijk and De Bruijn[28] are however of the opinion that the underrepresentation of the unemployed in volunteer work may to a considerable degree be described to level of education and occupational sector (see (2)).
This assumption is based on the fact that underrepresentation of the unemployed in volunteer work dates from a period when people with the level of education and belonging to the occupational group from which a larger percentage of volunteers derive, were less affected by unemployment.
Having a lot of free time is thus in itself no guarantee for particpatication in volunteer work. Educational level and occupational sector exert a greater influence on this phenomenon.
Moreover in the period 1975-1980 some alteration is to be observed in the way in which the unemployed and medically unfit spend their time: they now occupy their time relatively more actively.[31]

The trend toward greater participation in volunteer work will be further intensified since it is to be expected that the present obstacles to volunteer work[32] will be dispensed with.

The improved income position of the elderly in this scenario will also influence the choice of the type of facility. Surveys have shown that elderly persons with a low income are more inclined to appeal to subsidised facilities than the elderly with a higher income (assuming a similar state of health).[33]
More possibilities are open to people with higher incomes to make use of (commercial) aid 'cut to measure' (taxis, eating in restaurants, paid domestic aid). The expected improvement in the incomes of the elderly will consequently result in an increase of this help.

5.3.3 Nursing and medical treatment (diagnosis and therapy)

Nursing and medical diagnosis and therapy of the elderly take place in the intramural sector especially in nursing homes and hospitals. Nursing and medical treatment in the intramural sector also imply (partial) care and (semi)permanent sojourn. The extramural nursing function was already discussed in Section 5.3.2. Extramural medical treatment (especially by GPs) will be discussed in this section.

Shifts as a result of the changing concept of care will be of much less significance for the above-mentioned facilities than for residential and care facilities. This is mainly a consequence of the specialistic nature of these services, which can hardly be substituted in another manner. Nonetheless, there will be marginal changes. To give some examples:

(1) An increase in sheltered residential projects, emanating from nursing homes and homes for the elderly (see also Section 5.3.2).

(2) No further expansion of the so-called GAAZs (Geriatric Wards in General Hospitals). The reason for this is an undesirable shift in the nature of GAAZs in the direction of the nursing home. This however does not mean that geriatrics as a specialism will disappear. However, geriatrics will not be practised as a specialism with its own beds. The geriatrician will increasingly play a co-ordinating role in intra and extramural care for all the disciplines concerned with the care of the elderly.

(3) Making a start with setting up small-scale, decentralized long-stay facilities, for instance inspired by the Danish model. Here, the small-scale integrated home for the elderly/nursing home where treatment and residential functions are combined, is already a fact. Every occupant has his own appartment. This offers advantages from a point of view of privacy. Other advantages are that in the case of couples they do not need to be separated and there is no need for transfers from homes for the elderly to nursing homes.[34] Since, however - at least as far as policy is concerned (see Section 2.5) - there are few possibilities for expansion of nursing homes or homes for the elderly, the new type of long-stay facility will result mainly from adaptation of existing facilities. It will be evident that such a process can only come about slowly, certainly in the case of economic recession.

Computerization may exert a positive influence on the trends to reduction of scale.

Physicians in small units can for instance, thanks to an electronic network, communicate with physicians in more specialized, better equipped larger units for the purpose of making diagnosis and deciding upon therapy. By this means, advanced medical apparatus can be limited to a few centres.

(4) It will not be possible to bridle technological advances with respect to diagnosis and therapy (see Chapter 3). In this scenario there will however be only a selective use of these advances, primarily aimed at improving the quality of life of the older patient. Prolonging the process of dying will be avoided. Increasingly, early diagnosis will only be made if there is a reasonably good chance that therapy for treatment of the disease exists. If this is not so, early diagnosis might only serve to draw out the disease process, since the patient is sooner labelled sick.
As regards periodical large-scale preventative examinations of the elderly, costs and benefits will be weighed very carefully against each other.

(5) As far as GP help is concerned, the concept of scientific purity in medical practice (see central point of departure 4), gains ground as opposed to the physician as general solver of problems (medical, social and pastoral function of the physician).
This however does not detract from the fact that in the future the physician will increasingly practice his profession in a team with other disciplines.
A survey carried out by the Dutch Institute of General Practitioners[35] is worth mentioning in this context. This showed that GPs who work in a team (for instance in a health centre) refer approximately 10% less sick fund patients to specialists than solo GPs. Apparently a (multidisciplinary) team is sooner in a position to make a more complete diagnosis than the solo GP. Nonetheless other factors may play a role in the number of referrals (for instance the age structure of the practice).

(6) Another development is the phenomenon of treating oneself. This is made possible by the fact that in future simple (diagnostic) apparatus (for instance the sphygmomanometer) will become available to the public on a large scale. In addition to this, advances in medical protocol in combination with computerization will create possibilities for 'telediagnosis'. This will be an interesting development for less mobile older people. However it is likely that these methods will only be utilized on a large scale by generations of elderly people who have grown up in the computer era.

5.3.4 Conclusions with respect to facilities

In the pages which follow, we will attempt to provide some insight into the consequences of the developments discussed above for the pressure on facilities. If supply is in a position to react adequately to these social developments, this could in some instances lead to reduced increase or even decrease in the use of facilities by the elderly. Below, a distinction is made between (1) relative usage and (2) absolute usage.

(1) **Relative usage**

By this is meant usage related to the (increasing) size of the group of 65 years of age and older.

In the past, especially since the end of the seventies, the relative use of homes for the elderly, inter alia as a consequence of policy, has decreased. Relatively speaking, there has been no decrease in the use made of nursing homes, though there has been a levelling off of growth.

In this scenario it is expected that there will be an intensification of the tendencies described above since this scenario assumes that one combined nursing home and home for the elderly will come into being to which only patients in need of intensive assistance will be admitted. Though in this scenario there will be a relative decline in the use of this combined facility, the above probably implies an increase in costs per patient in view of the need for, inter alia, more staff.

In the recent past, relatively speaking, there has been a (slight) increase in the number of days spent in hospital by the elderly. Estimates of relative usage of hospitals by the elderly for the trend variant of the reference scenario show that a decrease may be expected. Since in the trend variant the factors from scenario C which could check growth are not taken into consideration, the relative use of hospitals by the elderly in scenario C might show a further decline.

As regards home help and district nursing, which in the past showed growth and stabilization respectively in the relative use made of these facilities, in this scenario an increased demand is expected since the reduced use made of homes for the elderly in this scenario will lead to a heightened demand for facilities of this type. The increase in alternative residential forms and in care and volunteer work will to some extent reduce the

increased demand, but will by no means constitute a complete alternative.
No changes are expected in this scenario in the sharp increase in day treatment at present observed.
In this scenario care by GPs and from ambulatory mental health care (AGGZ) will remain constant, partly in the light of the assumption of a - at least in the short term - constant state of health of the elderly.
There will be a relative increase in the use of adapted dwellings for independently residing elderly persons. Especially there will be an increase in demand for facilities for group households.

(2) **Absolute usage**
In cases where relative usage remains **stable** or **increases**, in view of the absolute increase in the number of elderly, absolute usage will also **increase**.
Where there is a **decline** in relative usage, this **by no means** implies a **decline** in absolute usage (cf. for instance homes for the elderly in the trend variant of the reference scenario.

In the above, a relative decline was assumed of the use made of hospitals and combined homes for the elderly and nursing homes. With respect to hospitals it is expected that this decrease will not be manifested in absolute terms. In this scenario an absolute decline in the use of the facility nursing home and home for the elderly is not considered improbable. The following reasons for this may be given. In Section 2.5.5 (homes for the elderly) it was stated that the number of beds in homes for the elderly in 1983 was 145,000. In this scenario there will be a reduction of beds in homes for the elderly as a consequence of the developments described, which might be consistent with a percentage of beds for people of 65 and older of 6% (compare norm: 7%). Six percent of the population of 65 years of age and older implies slightly more than 125,000 people in the year 2000. That is to say, in this scenario, in the sector homes for the elderly approximately 20,000 beds will be 'freed' which then become available for the long-stay facility for the elderly. These 20,000 beds will be more than enough to compensate the expected increase in the nursing home sector (cf. trend variant in reference scenario: an increase of 11,597 beds between 1983 and 1990; between 1990 and 2000 an increase of less than 11,597 on account of the tendency to levelling off of growth). As a result of this, there might be a decrease in the absolute use made of the combined facility home for the elderly and nursing home.

Naturally, this will in turn have consequences for other facilities.

Finally, we would like to devote some attention to the costs aspect. The decrease or increase of expenditure for facilities can be explained by three factors:

- shifts in relative use, apart from demographic developments;
- demographic developments;
- shifts in costs per unit of use (increase in costs per bed, for instance).

The SCP[36] has made expenditure estimates covering various facilities for the elderly on the basis of these three factors. The estimates show that in the past the third factor for the most part accounted for rises in expenditure. The increase in costs till the year 2000 as a consequence of the combined effect on the three factors by comparison with the base year 1981 are 19% for homes for the elderly and 36% for nursing homes respectively.[37]
Though there may sometimes be a decrease in costs,[38] generally speaking it is not to be expected that costs will decline. That costs will be reduced by substituting extramural care for intramural care is only an assumption which has not (yet) been borne out by empirical data.[39] Table 5.4 gives an overview of the above.

Table 5.4 Expectations with respect to usage of subsidized professional facilities in scenario C

	relative use	absolute use
combined nursing home/home for the elderly	-	-
day treatment	+	+
district nursing	+	+
home help	+	+
(adapted) dwellings for independent (group) occupancy	+	strong +
hospitals	-	slight +
GPs	0	slight +
AGGZ	0	slight +

See notes 5 and 6 in Chapter 1 for an explanation of the signs and the concept 'relative use'.

5.4 The economic context

In this section we discuss the question how assumptions concerning economic growth or shrinkage relate to the scenario presented here.

Economic growth is not necessarily a condition for this scenario, but it would nonetheless serve to modify the effects of a possible reversal in income resulting from shortening of working hours. Furthermore, in the case of economic growth no drastic alterations are expected in the consequences for facilities by comparison with the situation described in Section 5.3, since this scenario is not so much based on financial motives but rather on more intrincic motives.

If the shortening of working hours should not have the desired effects in combating unemployment, a development in the direction of a basic income for everyone would appear to be not unlikely.
The basis income would take the place of all grants and everyone would be free to supplement it to the extent they desire from payed work. With this, there would cease to be an unequivocal link between work and pay.
Unemployed elderly persons will then be a phenomenon just as often encountered as unemployed persons in other age categories. This would mean that in this respect the elderly would be on the same footing with other age categories.

The levelling off of (the growth of) demand for facilities described in this scenario would be opportune in a situation of economic decline. If economic deterioration necessitates an even further shrinkage in the (growth of) demand for facilities, solutions will be sought in the sphere of for instance (limited) compulsory service for the youth and an increase of 'unconditional income transfers' for the elderly. As opposed to 'conditional income transfers' (for instance rent subsidies, subsidies for care facilities), 'unconditional income transfers' (old age pensions, for instance) enlarge the expendable income of the elderly. As a result of this, as was already pointed out in Section 5.3.2, the demand for subsidized professional care will be reduced. Though the increase of unconditional income transfers will be financed from public funds, this will probably be more than compensated by the reduction in the use of facilities.
The reader is referred to Table 1.1 in Chapter 1 for a complete overview of the shrinkage scenario.

Notes Chapter 5

(1) Terminology derived from J. MacKnight. Quoted by Robertson, 1983, p. 19.

(2) Harmonization Council for Welfare Policy (HRWB), 1983, pp. 7-11.

(3) Quoted by the Social and Cultural Planning Bureau (SCP). 1984, p. 65.

(4) SCP, 1984, p. 65.

(5) SCP, 1984, pp. 11 and 277.

(6) De Kam and Van Tulder, 1983.

(7) Expectations of Verwey-Jonker: Between 2000 and 2010, people now retired prematurely on a voluntary basis will, in the absence of younger people, make way for those who would like to continue to work after the age of 65.

(8) Society of Actuaries, 1983.

(9) Scientific Council for Government Policy (WRR), 1982, pp. 193.

(10) Havighurst, 1978, p. 23.

(11) In view of their limited importance, no account has been taken of incomes deriving from property.

(12) Katz et al., 1983, p. 1219.

(13) Van der Maas, 1982.

(14) This in spite of the contradictory example of Russia (Section 2.3). In Russia however, other factors may be responsible for the large gap in life expectancy between the sexes.

(15) Lägergren et al., 1984, p. 8.

(16) Dooghe and Bruynooghe, 1984, p. 91.

(17) Cf. the title of the publication by Lägergren et al., 1984.

(18) Van der Zanden, 1982.

(19) Hospitals Council of the Netherlands (NZR), 1981.

(20) Harmonization Council for Welfare Policy (HRWB), 1983, p. 18.

(21) Houben, 1984.

(22) Weeda, 1982, p. 100.

(23) SCP, 1984, p. 302.

(24) Weeda, 1982, pp. 87 and 93.

(25) Compare the experiment with slightly dementing elderly patients of the nursing home De Landrijt, Eindhoven.

(26) Knipscheer, 1984, p. 83.

(27) CBS, 1984.

(28) Van Luijk and De Bruijn, 1984, pp. 143-144.

(29) SCP, 1984, pp. 66 and 359.

(30) WRR, 1982, p. 181.

(31) SCP, 1982, p. 56.

(32) Compare the reports of the Interdepartmental Committee on Volunteer Policy, 1980, 1981, 1982.

(33) Pommer and Wiebrens, 1984, p. 80.

(34) Biesenbeek, 1983, pp. 22-24.

(35) Hospitals Institute of the Netherlands, 1983.

(36) Goudriaan et al., 1984.

(37) Goudriaan et al., 1984.

(38) Cf. an experiment with sheltered dwellings for slightly dementing patients of De Landrijt in Eindhoven, where a reduction of costs of 60% was achieved by comparison with sojourn in a nursing home. Critics considered that the reduction had been calculated too favourably, since factors such as living in a subsidized dwellings etc. were not taken into account in the calculations.

(39) Van Santvoort, 1984, pp. 32-33.

6 Disturbing developments

6.1 Introduction

In this chapter two disturbing developments are discussed, namely:

(1) A five-year postponement of dementia.
(2) An extreme reduction in intergenerational solidarity (care of elderly parents by children).

The developments chosen for discussion are very different in character: the former (postponement of dementia) relates to medical and medical-technological developments, and might perhaps reduce the demand for (health) care facilities, while the latter relates to social developments and will undoubtedly result in an increased demand for professional aid. This will be discussed in greater detail in the concluding section (6.4.).

6.2 Postponement of dementia

6.2.1 Justification of choice

A number of reasons may be given for choosing dementia as the subject of one of the disturbing developments, namely:

- Dementia is a complex of symptoms mainly encountered among the elderly, and is consequently very relevant in the context of scenarios on ageing. Dementia manifests itself in many ways. For this reason the concept **Chronic Brain Syndrome** is often employed. For the sake of brevity, in the ensuing pages we will however speak of dementia.
- Dementia results in a high degree of impairment and tarnishing of the personality.
- As is known from epidemiological literature, incidence and prevalence of dementia increase with age. In the light of the increasing (double) ageing of the Dutch population, this means an increase of both the absolute and the relative share of demented people in the age category 65 years of age and older.

For this reason dementia is sometimes referred to as 'the epidemic of the century' or 'the silent epidemic'.

- The increasing number of dementing old people also has consequences for facilities. Demented people stay relatively more often in institutions. Whereas, generally speaking, the ratio of elderly people who live at home to those in institutions is 90% to 10%, in the case of dementing old people the ratio is estimated at 75% to 25%.[1]
- Though dementia in the Netherlands is not registered as a primary cause of death, according to some authors it should nonetheless be viewed as a major killer. In any case, epidemiological literature shows that the life expectancy of demented patients is in general lower than for the elderly as a whole.

In Section 6.2.2 first of all an overview is given of some epidemiological aspects of dementia. Section 6.2.3 deals with medical and medical-technological developments relating to dementia in as far as these have become evident from the literature and group discussions. In Section 6.2.4 the results of calculations with respect to a five-year postponement of dementia are given. In view of the limited availability of research data, calculations relate only to the nursing home sector.

6.2.2 Some epidemiological aspects of dementia

As we already said, dementia is not a disease with clearly defined symptoms. The problem is not solved by substituting such terms as 'brain failure' or 'chronic organic brain syndrome' as is done by some authors. One advantage of these two terms might be that they are less stigmatizing than the term dementia.[2]

The problem surrounding the term dementia also reflect upon the epidemiological data on dementia. Figures with respect to the prevalence and incidence of dementia among the elderly vary considerably from country to country. Ringoir[3] notes that prevalence figures vary from 0.5 to 31.8%. In the Netherlands there is the further complication that hardly any epidemiological data are available, so that it is necessary to fall back on data from other countries. In addition to the limited agreement as to what consitutes dementia, other factors too contribute to the divergence of epidemiological data. Contributing factors are:

(a) Differences in research methods used:
   e.g. the use of a questionnaire with a check-list of symptoms by non-professional interviewers versus diagnosis by physicians.
   e.g. studies based on patients' contacts with existing institutions (psychiatric case registers) versus field studies.

(b) Differences in age limits. Some studies relate to people of 60 and older, while others employ the age limit 65 or 70.

(c) Differences in age grouping. Age grouping among the persons surveyed is important since prevalence and incidence show a progressive increase with age. If there are more (very) old people in one survey group than in another, even if the age specific figures are similar, this will lead to a difference in the **overall** prevalence and incidence figures for the total group of elderly.

(d) Differences in distribution over the sexes. If the distribution according to sex differs in two studies, this might result in a difference in the overall prevalence and incidence figures for the elderly.

The above will have made clear that calculations should thus as far as possible proceed from age specific figures (per five-year cohort) and sex-specific figures.

In the Netherlands, data on the prevalence of dementia and psycho-geriatric problems are mainly limited to the residents of **the intramural sector**. A survey carried out in homes for the elderly in the Almelo region in 1980 showed that in the opinion of personnel, 4.7% of the residents were not suited for such homes in view of psychic disfunctioning. Where other criteria are employed (for instance the Assessment Scale for Assistance Requirements of Elderly Patients (B.O.P.) or the scale employed by the Hospitals Council of the Netherlands), approximately 17% of the inmates of homes for the elderly are found to be psychically handicapped.[4] A survey in Friesland dating from 1969 on the other hand showed psycho-geriatric problems among 27% of the residents of homes for the elderly.[5]
For our purposes however, the most interesting data are those concerning the prevalence of dementia, collected since 1981 by the Nursing Home Information System of the Information Centre for Health Care (SIVIS). Use will be made of these data when making calculations on the postponement of dementia (see 6.2.4).

It is in any case worthwhile to have a general overview of the prevalence of dementia. For this purpose, use is made of so-called 'pooled data'. Prevalence data deriving from various epidemiological studies on dementia in other countries have been combined (see Table 6.1).

Table 6.1 Prevalence of dementia according to sex and age in percentages

age	men (N=1008)	women (N=1259)	combined sexes (N=2267)
65 - 69	3.9	0.5	2.1
70 - 74	4.1	2.7	3.3
75 - 79	8.0	7.9	8.0
80 +	13.2	20.9	17.7
all ages	6.2	6.3	6.3

Source: Kay and Bergmann, 1980, p. 43.

The data relate to the 'chronic brain syndrome'. According to Kay and Bergmann, this includes two important forms of dementia, namely:

(1) Senile dementia of the Alzheimer Type (SDAT), and
(2) Vascular (or multi-infarct or arteriosclerotic) dementia.

Estimated at 50% of all cases of dementia, SDAT is the most often encountered form of irreversible dementia. Vascular dementia accounts for between 8 and 20% of cases of irreversible dementia.6)

The following conclusions with respect to the ratio between the two types of dementia mentioned may be drawn from studies pooled by Kay and Bergmann:

(1) Under the age of 75 arteriosclerotic dementia is encountered among men more often than SDAT, while both types are rare among women under the age of 75.
(2) Over the age of 75, SDAT is just as prevalent or even more prevalent among both sexes than arteriosclerotic dementia.

(3) SDAT is a disease mainly encountered among people over the age of 75; this also applies to arteriosclerotic dementia, at least among women. Among men it becomes manifest at an earlier age.

If dementia is classified according to severity, the figures given in Table 6.1 relate only to the categories 'severe' and 'moderate'. To these two categories accrue people who are no longer capable of looking after themselves: in other words, potential inmates of institutions.[7] The category 'mild' dementia thus falls outside the limits of Table 6.1.

If the prevalence data from Table 6.1 (age and sex-specific) are applied to the Dutch population of 65 and older in 1983, the following figures would be obtained (see Appendix J for calculations).

Table 6.2 Estimated number of dementing persons (exclusive of mild cases) in the Dutch population of 65 and older in 1983 (absolute and relative)

sex	absolute (to the nearest thousand)	%
men	44,000	6.4
women	73,000	7.8
total men + women	117,000	6,9

It is a known fact that mortality among dementing old people is higher than among non-dementing old people. The rule of thumb adhered to is that the life expectancy for dementing old people is 0.75 of the life expectancy of other elderly persons of comparable age and sex. Boelen and Zwanikken[8] pose that on the average senile dementia results in death in five years.
Though dementia (and especially SDAT) may thus be viewed as a 'major killer' which might well be listed fourth or fifth in the table of causes of death, in the Netherlands dementia is never registered as a primary cause of death, so that dementia patients are classified in another category of causes of death.[9]

6.2.3 Medical-technological developments relating to dementia

In the pages which follow, medical and medical-technological developments relevant for dementia which may be expected in the (near) future are discussed as these emerged from group discussions and to a certain degree from the literature (see also Chapter 3 and Appendix I).

Generally speaking, it may be posed that in the coming years till the year 2000, no breakthroughs of any significance may be expected in research as to the causes and etiology of dementia. One of the reasons for this is the absence of an adequate animal experimental model. The suggestion has however been put forward that in future not one, but various models for the various aspects of dementia will be sought.[10)]

As regards therapeutic potentials relating to dementia, it is expected that for the future, research will continue among much the same lines as at present. These include:

- Research as to neuropeptides. It is possible that neuropeptides might serve to lessen the loss of memory with which dementia is accompanied so that this promises a certain 'resocialization' of demented persons.
- Research of transmitter normalization. As in the case of Parkinson's disease, attempts have been made to treat SDAT by supplementing neurotransmitters, most attention being devoted to the cholinergic system. Clinical trials in this area have however up to now been disappointing, possibly on account of the fact that as yet not all neurotransmitters are known.[11)]
- In general, little hope is entertained with respect to other pharmacological therapies. Little is expected of drugs which were not designed specifically for psychogeriatric purposes such as the classic psychopharmaca (antidepressants for instance), since they work symptomatically and are not effective in the case of dementia. As regards to the specific gerontopsychiatric drugs, there has been no convincing evidence that they halt or reverse the process of dementia.[11)] These drugs often function as placebos.
- In addition to the various therapies in the area of pharmacology mentioned above, non-pharmacological therapies increasingly call for attention. There are for instance indications that it is possible to exert a positive influence on dementia by adapting the patient's (social) environment. A number of therapies may be

listed under this denominator, as for instance:

- 'structured stimulation', a therapy based on the idea that some symptoms of dementia may be ascribed to sensorial deprivation.
- the 'Reality Orientation Therapy' (ROT), by means of which it might be possible to postpone or prevent mental deterioration.
- 'the enriched environment'. It is well known that an 'enriched environment' stimulates brain development and limits the deficits resulting from developmental disturbances. Even in adulthood such an environment can stimulate the formation of new dendritic branches.

Other names for therapies of this nature are sociotherapy or environmental therapy.
Another non-pharmacological therapy might perhaps be sought in altering the dietary pattern of the elderly (inter alia the eradication of vitamin deficiencies).

- Finally we should not omit to mention possibilities for the transplantation of embryonic brain tissue. Many problems, including ethical ones, will however have to be solved before this is possible.

For the coming years, prospects with respect to diagnosis of dementia (especially early and differential diagnosis) are more promising than with respect to etiology and therapy. Improved diagnosis would be made possible by:

- More attention in the training of physicians (GPs) for diseases relating to age, including dementia.
- Better counselling of the public in general, as a result of which dementia would be sooner recognized by the patient or by people with whom he or she has contact (a partner).
- Improved technological possibilities in the form of scanners. In this connection mention should be made of PET scan (Positron Emission Tomography), with which the functioning of the brain can be studied, and NMR (Nuclear Magnetic Resonance) with which it is possible to study the structure of the brain. With the aid of these techniques, it would be possible to distinguish forms of dementia which might react to treatment.
  It is possible that the PET Scan might also be capable of revealing positive signs of SDAT. At the moment diagnosis is based on a process of elimination.[12] Indeed, technological innovations in the area of diagnosis should go hand in hand with developments of psychological/psychiatric instruments for the making of diagnosis (behavioural observation scales, for instance the Assessment Scale for Older Patients (BOP)).

- Increasing multidisciplinary co-operation in the observation of patients suspected of being demented. Reference might be made to the multidisciplinary screening teams which already operate in various parts of the country. Such screening teams offer good chances for distinguishing reversible (pseudo)dementias from irreversible forms. For instance Claessens[13)] in his recent doctoral thesis points out that of 230 patients recommended for admission to a psychogeriatric nursing home, 71 (30.9%) were found to be suffering from a reversible form of dementia. Similar conclusions may be drawn from other surveys carried out in the Netherlands.[14)] Among these reversible forms, drug intoxication may be found to play a role.
- Further expansion or intensification of ambulatory mental health care (RIAGG and RIGG).

The importance of early diagnosis is that if dementia is diagnosed before the patient's condition has deteriorated severely, possibilities might exist for postponing some forms of dementia for some years.

6.2.4 Towards a calculation model

In view of the above, it is extremely difficult to calculate what will be the consequence of the developments mentioned for the prevalence of dementia and the need for facilities for these patients till the year 2000. Should it be assumed that the developments are of marginal value and will have hardly any impact on the prevalence of dementia before the year 2000? Or would it be reasonable to expect some positive influences on the prevalence of dementia? The benefits to be expected would rest mainly in possibilities for postponing dementia for a number of years thanks to better diagnosis (and possibly some therapeutic measures).
In this section a rough estimate will be made of what the consequences for the **nursing home sector** would be if it should prove possible to achieve a five-year postponement of dementia. The results will be viewed against the background of the null situation, the situation in which no changes will come about. Limitation of this estimate to the **nursing home sector** is determined by the earlier-mentioned scarcity of data. Naturally, a lot of demented patients are nursed at home, while others are in homes for the elderly. They may or may not be assisted by semimural facilities (day-treatment, treatment in outpatient departments) and extramural care (GP, district nurse, RIAGG).

Demented old people are also admitted to general and psychiatric hospitals. In these sectors too a five-year postponement would have consequences for the care needed. However - on the basis of costs[15] - the nursing home sector constitutes the most important form of care.

Some key data on patients in nursing homes with the primary diagnosis dementia,[16] are given in Table 6.3.

Table 6.3 A few key data on patients of 65 and older in nursing homes with the primary diagnosis dementia

	1981	1982	1983
number of patients on 31.12	11,661	11,779	12,494
number of (re)admissions	4,266	4,137	4,036
number of discharges/deaths	4,097	4,140	4,015
average number of days in home per discharged/deceased person	847,51	906,33	918,46
total number of days spent in homes by patients discharged or deceased	3,472,268	3,752,189	3,687,622

Source: SIVIS

The reader is referred to Appendix K for the calculation method employed.

A five-year postponement of dementia can be interpreted in more than one manner. In the pages which follow, we will proceed from the premise that in the future the age-specific admission figures for dementia will shift forward by one five-year cohort. In other words: in the future the age-specific admission figure in the age category 65-69 will be equal to the present age-specific admission figure for the age category 60-64, etc.

It should be pointed out that the path to carrying out such calculations is strewn with so many pitfalls that the end result should be seen rather as an illustration of the problems with which one is confronted than a realistic estimate of the future situation.

Especially it might be questioned how realistic are the assumptions which have been used for the purpose of simplifying the calculation model. For instance, the model proceeds from a 'steady state', while in fact the situation shows constant change (see e.g. Table 6.3). Moreover, certain factors are assumed to remain unchanged, while there is no way of telling whether they will indeed remain the same in the future (for instance an unchanged policy with respect to admission to nursing homes, the same willingness to keep a demented person at home, etc.).
The results of the calculation - the value of which should thus be considered limited - are given in Table 6.4. This Table shows the situation which would arise if in the year 2000 the postponement of dementia by five years had become a reality. The effect of this postponement would of course continue after 2000, but no calculations have been made after this date.

Table 6.4 shows that in the year 2000 there would be approximately 26% less demented patients in nursing homes if there was a postponement of dementia than if there was no postponement. The difference is somewhat greater among women (28%) than among men (21%). By comparison with the situation at the beginning of the eighties, realization of a postponement of dementia in the year 2000 would mean an increase of 9% in the number of dementing old people (men 4% and women 10%). If there is no postponement of dementia, the number of dementing people in nursing homes would increase by 48% by comparison with the situation at the beginning of the eighties (men 33% and women 52%).

Table 6.4 Number of patients of 65 and older in nursing homes with the primary diagnosis dementia in the period 1981-1983 and 2000 (with and without postponement) according to sex

	average	without postponement	with postponement
men	2,699	3,587	2,816
women	9,279	14,144	10,225
total	11,978	17,731	13,041

In conclusion, we refer to Section 6.4 which contains a number of remarks with respect to the making of such a calculation, and warnings are given as to being too optimistic as to the decreased pressure on nursing homes resulting from a postponement of dementia.

6.3 Extreme decrease of intergenerational solidarity

6.3.1 Justification of choice

In scenarios B and C some decrease of intergenerational solidarity was already assumed. In the disturbing development discussed in this section, the consequences are considered of the fairly extreme situation which would come about if in the year 1990 or 2000 there should be a sudden and **complete** halt to the aid with which children provide their parents. In reality, it is unlikely that such an extreme situation will come about, but this calculation is only intended as an indication.

There are a number of reasons why it is not relevant to consider the possibility that in the future there will be a decrease in the amount of umbrella care with which children provide their parents. These include:

- The decreased number of children and the fact that children have long led an independent life by the time their parents require assistance. Moreover, by this time the children themselves are already advancing in years.
- Increasing emancipation of women as a result of which the willingness and ability of daughters (in law) to provide assistance to parents (in law) may decrease.
- An increasing tendency to individualization.
- Increasing geographic mobility, which often means that the distance between parents and children is too great for children to provide assistance.
- The increasing disinclination on the part of the elderly to accept the asymmetric aid relationship which comes about in the case of informal aid (inter alia from children). If symmetric assistance **(reciprocal aid)** is no longer possible, the elderly prefer to call in professional aid.

The data for these estimates were derived from the Survey of Living Conditions carried out by the CBS in 1976 among the Dutch population of 55 and older.

6.3.2 Some basic data on the aid relationship children-parents

In the 'Living Conditions Survey' a distinction was made between elderly people living independently and elderly persons in homes. The latter category of elderly will not be included in the calculation model, since in this case children are not called upon to make any contribution in the form of aid. In the pages which follow, the relationship between independently residing elderly persons and their children is confined to children living **away from the parental home** and their parents. Children living at home have not been included in the calculations since the survey provides no indications as to why children continue to live with their parents. Their presence does not necessarily denote the provision of aid.

Table 6.5 shows what percentage of independently residing elderly persons fairly regularly receive aid from children living away from home in the performance of ADL and/or domestic tasks (column 2).

As Table 6.5 shows, elderly people's own children who no longer live at home as well as paid domestic aid are numerically the most important sources of aid.

The living conditions survey also showed that assistance from children living away from home is considerably greater when the parents live alone than when they live with others (a partner, other children).

Table 6.5 Percentage of independently residing elderly persons who (fairly) regularly receive assistance from various categories of persons residing elsewhere in the performance of ADL and/or domestic tasks according to age and sex in 1976

	children	other family	neighb., friends, volunt.	payed comm. help	homehelp/ spec.for elderly	distr. nurse	100% abs.
men							
55-59	3.1	2.2	2.6	7.3	1.2	0.4	509
60-64	4.9	2.3	3.0	7.9	1.4	0.7	430
55-64	3.9						
65-69	7.2	2.9	3.6	7.4	2.5	0.9	443
70-74	12.3	2.3	4.5	8.8	6.3	1.5	397
75-79	24.3	4.0	9.5	12.4	15.3	4.0	346
80 +	22.0	9.8	10.5	19.7	15.3	6.8	295
tot. $\geq$ 55	8.8	3.1	4.3	9.0	4.4	1.5	2420
tot. $\geq$ 65	13.7	3.9	5.9	10.4	7.5	2.4	1481
women							
55-59	4.9	1.8	3.9	8.3	2.2	0.8	508
60-64	7.2	3.0	3.3	10.9	2.4	0.7	541
55-64	6.1						
65-69	13.2	3.5	6.0	11.7	4.2	1.1	546
70-74	13.2	5.8	7.2	15.8	7.5	2.1	530
75-79	23.8	7.7	12.6	16.7	11.0	2.4	508
80 +	25.9	8.8	13.9	21.4	17.0	8.5	294
tot. $\geq$ 55	11.7	4.1	6.2	12.5	5.3	1.6	2927
tot. $\geq$ 65	16.7	5.6	8.6	15.0	8.0	2.5	1878
men+women							
55-59	4.0	2.0	3.2	7.8	1.7	0.6	1017
60-64	6.2	2.7	3.2	9.6	2.0	0.7	971
55-64	5.1						
65-69	10.5	3.2	5.0	9.8	3.4	1.0	989
70-74	12.8	4.3	6.0	12.8	7.0	1.8	927
75-79	24.0	6.2	11.4	15.0	12.6	3.0	854
80 +	23.9	9.3	12.2	20.5	16.1	7.6	589
tot. $\geq$ 55	10.3	3.6	5.4	10.9	4.9	1.6	5347
tot. $\geq$ 65	15.4	4.9	7.4	13.0	7.8	2.4	3359

Source: CBS, 1983, p. 62.

Some insight could also be obtained from the living conditions survey into the amount of help provided for ageing parents by children living elsewhere. Of children who assist their parents, in 67% of cases aid is provided for 1-6 hours per week and in 18.9% in cases for more than 6 hours (see Table 6.6).

Table 6.6 Independently residing elderly people of 55 years of age and older who (fairly) often receive assistance with ADL or other domestic tasks from persons living elsewhere according to average number of hours assistance per week in 1976

	1-6 hrs.	more than 6 hrs.	no. hrs. unknown	total	abs.
children	67.0%	18.9%	14.1%	100%	670
other family	65.7%	17.4%	16.8%	100%	226
neighbours/ acquant./ volunteers	73.7%	10.6%	15.7%	100%	338
commercial help	66.5%	26.5%	6.8%	100%	637
home help/ specially for elderly	53.9%	41.5%	4.7%	100%	338
district nurse	82.1%	8.1%	9.7%	100%	111

Source: CBS, 1983, p. 71.

The above data show clearly that aid to independently residing elderly persons does not derive primarily from those sources which are most characteristic of the welfare state, namely home help and district nursing. Informal aid, especially from their own children, continues to play an important role. In addition to this, commercial domestic aid, usually financed completely by the elderly person, is also of considerable significance.

6.3.3 Towards a calculation model

Calculations proceed from the assumption that all other conditions except the umbrella care provided by children will remain constant. First of all, for the purpose of the calculations it is essential to determine the number of independently residing elderly persons in the years 1990 and 2000. If the same percentage of institutionalization in homes for the elderly and nursing homes as at the beginning of the eighties is assumed, the number of independently residing elderly can be calculated (age and sex-specific) by subtracting the number of institutionalized persons from the total forecasted number of elderly in 1990 and 2000. With the aid of the percentages given in Table 6.5, it can then be calculated how many of the independently residing elderly in 1990 and 2000 (according to age and sex), receive aid from children living elsewhere.
The calculations are given in Appendix L. The results of these calculations are shown in Table 6.7.

Table 6.7 Estimated absolute number of independently residing elderly persons of 55 and older who receive aid from children living elsewhere in 1990 and 2000 according to sex (to the nearest thousand)

	1990	2000
men	124,000	142,000
women	216,000	238,000
total men and women	340,000	380,000

If children should cease to provide their elderly parents with umbrella care, this would mean that in 1990 and 2000 340,000 and 380,000 persons respectively would have to seek this essential aid elsewhere. It can only be speculated in what direction and to what degree the demand for help will develop. Below, one possible example is given of the manner in which the demand for care resulting from the failing off of umbrella care by children might be met. There are of course numerous other possibilities. Supposing we proceed from the following assumptions:[19)]

- In cases where less than 6 hours of aid is provided (67%, see Table 6.6) it is assumed that this will be supplemented by first echelon facilities and a paid domestic help (in the ratio 50%:40%). In addition to this it is assumed that in 10% of cases removal to an adapted dwelling will be sufficient.
- In cases where children living elsewhere provide more than 6 hours of aid per week (18.9%), it is assumed that 50% of elderly people move to homes for the elderly, 40% will appeal to first echelon facilities, while only 10% will call in commercial aid (in view of the large number of hours involved, commercial domestic help is costly).
- With regard to 14.1% of the elderly it is not known how many hours of assistance they receive from children. It is assumed that half of these elderly persons will appeal to first echelon facilities while the other half will call in commercial aid.

For the sake of simplicity, it is assumed that the various types of facilities preclude each other. This is however not necessarily the case with respect to all facilities. According to this example, some of the elderly will thus be able to remain independently at home by, either for the first time or to a greater degree than formerly, appealing to first echelon facilities or commercial aid, while the rest of the elderly will move to homes.

Appendix M shows the calculations for the years 1990 and 2000 on the basis of the assumptions discussed. The sum total of these developments will be the following increased demand on facilities as a result of the disappearance of intergenerational solidarity (Table 6.8).

Table 6.8 Estimated increase of the appeal to facilities in number of persons as a result of the disappearance of intergenerational solidarity with children living elsewhere in 1990 and 2000 (to the nearest thousand)

	1990	2000
homes for the elderly	32,000	36,000
adapted dwellings	23,000	25,000
first echelon facilities (district nursing and home help)	164,000	183,000
commercial aid	122,000	136,000

Other sets of assumptions will yield other results. If for instance it is assumed that in cases where children provide less than six hours of aid (67%), the elderly themselves undertake these tasks, the demands made on facilities would be considerably less than according to the estimates in Table 6.8. The reader can easily calculate the consequences of these and other variants.

6.4 Intersecting developments in relation to the scenarios

In the above it was calculated what a possible five-year postponement of dementia would mean for the prevalence of dementia in nursing homes.
We already pointed out that the conclusions drawn were based on assumptions, the probability of which should be further investigated.
Below, three reasons will be briefly discussed why one should be on his guard against overestimating the value of such experimental calculations.

(1) In the first place it should be pointed out that at the moment there is a (large) shortage of psychogeriatric beds. The norm for psychogeriatric beds has not yet been reached (1.25% of the number of persons above the age of 65).
According to this norm the number of psychogeriatric beds in 1983 should be 21,150, while in fact there were only 19,614. Moreover, the norm of 1.25% is fairly arbitrary. At the moment, the question of raising the norm to 1.5% is under discussion, also in view of the fact that studies carried out in other countries have shown that approximately 6% of 65-plussers suffer from a degree of dementia which is sufficiently severe to make them incapable of looking after themselves (see Table 6.1 with explanation). Thus, even if there should be a five-year postponement of dementia, there would still be a great demand for nursing home beds for this category of persons.

(2) The second reason is the multiple pathology encountered among the elderly. Data from the SIVIS show that multiple pathology is also regularly encountered among dementing patients. Table 6.9 shows the **number** of secondary diagnoses among patients whose primary diagnosis is dementia.

Table 6.9 Number of secondary diagnoses among patients with the primary diagnosis dementia (1981) (all ages)

Number of secondary diagnoses	Number of patients	Percentage
0	3,472	29.6
1	3,862	33.0
2	2,491	21.3
3	1,108	9.5
4	507	4.3
5	239	2.0
6	36	0.3
7	5	0.0
total	11,720	100,0

Source: SIVIS

Figure VI.1 gives an overview of the nature of the secondary diagnoses among patients whose primary diagnoses is dementia. The ten most often encountered secondary diagnoses among dementing patients are listed.

Figure VI.1 The ten most often encountered secondary diagnoses among patients whith the primary diagnosis dementia (1983)

- diabetes mellitus (adult onset or non-specified form)
- gonarthrosis
- coxitis
- heart block and dysrhythmia
- cardiac decompensation with back pressure
- degenerative and vascular disturbances of the organs of hearing
- paralysis agitans
- cataract
- delayed consequences of cerebrovascular diseases
- other cardiac complaints[20]

Source: SIVIS

The above shows that though as regards the primary diagnosis the psychogeriatric patient is clearly distinguishable from the somatic patient, if account is taken of secondary diagnoses, the psychogeriatric patient evinces a number of somatic complaints which are equally characteristic for the primary diagnoses of somatic patients. On the basis of these secondary diagnoses it is quite likely that this group of patients would continue to be eligible for admission to homes, even in the case of postponement of dementia.

(3) Thirdly, as was said earlier, a postponement of dementia will probably result in a prolonging of life. This will in turn imply that in the years gained the persons in question will run a heightened risk of contracting other diseases with the related consequences for (health) care facilities.

The disappearance of umbrella care by children would have the following consequences for the scenarios. If data from Table 6.8 are compared with data relating to demand for facilities in the reference scenario, this means that for homes for the elderly for instance, the disappearance of umbrella care by children in 1990 will mean a percentage increase of 22% by comparison with the trend variant and an increase of 19% by comparison with the null variant.

The disappearance of intergenerational solidarity will have less (dramatic) consequences for scenarios B and C since these scenarios already take account of a certain decrease of this solidarity. Nonetheless, in scenario B this disturbing development will mean extra pressure on professional facilities.
In scenario C this development may have the effect of halting the decrease in the (growth of) demand for facilities.

Notes Chapter 6

(1) Kay and Bergmann, 1980, p. 44.

(2) Ringoir, 1981, p. 31.

(3) Ringoir, 1981, p. 34.

(4) Fuldauer and Langendijk, 1980, pp. 3 and 8.

(5) Luteijn et al., 1972.

(6) Fliers, Lisei and Swaab, 1983b, p. 13.

(7) Kay and Bergmann, 1980, pp. 41 and 43.

(8) Boelen and Zwanikken, 1975, p. 66.

(9) Katzman, 1976, quoted by Ringoir, 1981, p. 36.

(10) Katzman, 1978, quoted by Goedhard, 1981.

(11) Fliers, Lisei and Swaab, 1983a, pp. 15 and 16.

(12) Fliers, Lisei and Swaab, 1983a, pp. 7 and 22.

(13) Claessens, 1984, p. 89.

(14) The studies by Miesen (1979), Fuldauer et al. (1980), Van Wersch-Van der Spek (1980) and Schouten (1979). Quoted by Claessens, 1984, p. 114.

(15) Ringoir and Van Duuren, 1981: Nursing homes f 808.8 million (1978). For comparison: Day treatment: f 2 million and extramural care f 274 million.

(16) Codes 210 and 212 of SIVIS, corresponding with senile dementia (ICD 290.0, 290.2, 290.3) and dementia resulting from multiple infarcts of the brain inclusive of arterio-sclerotic dementia (ICD 290.4) respectively.

(17) Re-admissions are only counted if the patient was a new case of 210 or 212 (SIVIS code).

(18) Meanwhile, this survey has been repeated in 1982. However, the results have not yet been fully published. Another reason for continuing to use the data from 1976 is that this survey includes aid with ADL or domestic duties, which are of primary importance for our purposes. The 1982 survey (only) deals with assistance in a more general sense.

(19) It is assumed that in instances where children living away from the parental home cease to provide assistance, the elderly persons concerned are in any case not eligible for admission to a nursing home. If the aid were provided by children living in the parental home, this might be the case.

(20) This refers to all heart complaints except acute myocardial infarct, angina pectoris, chronic pulmonocardiac complaints, acute and subacute endocarditis, valvular disorders of the heart, total atrioventricular block, other cases of heart block and dysrythmias of the heart, cardiac decompensation with back pressure and left decompensation.

7 Application possibilities for scenarios

7.1 Introduction

Scenario reports yield the most value in the hands of active readers. An active reader does not only wish to see overviews and conclusions, but would himself like to go to work with the scenarios. This can be done by means of mental experiments, supplementing the scenarios, carrying out additional calculations, in short using the scenarios as tools. A lot of readers will have to carry out these activities with scenarios themselves, while others, for instance policy executives, will have the necessary staff for this work.

Section 7.2 discusses more generally how scenarios can be employed. This section further clarifies what is meant by scenarios as 'learning environment' for policy preparation and implementation. It also offers insights into how the results of the learning process can be put to use in the realization of change.

Next, in this chapter the three scenarios from earlier chapters are confronted with patterns of health care facilities for the elderly. In Section 7.3 we see the conclusions to which such a confrontation can lead.

This chapter deals only with some of the possibilities which the reader might employ when going to work with the scenario report on ageing. That in addition to the examples given there are other possibilities for using the scenarios is discussed in Section 7.4.

In the concluding remarks (Section 7.5), an effort is made to derive some conclusions from the scenario project by confronting the results with the research questions defined in Section 1.2.

7.2 Scenarios as 'learning environments'

New policy can result in disappointments. Unforeseen circumstances may conflict with policy resolves. The effects may prove to be other than were anticipated during the preparation of policy. In order to guard against such disappointments as far as possible, researchers have sought for 'test situations' for policy resolves. In some countries it is already customary to test new policy beforehand. Think for instance of 'technology assessment'.
Anyone wanting to test new policy must imagine situations under which the policy would later have to be carried out. In so doing it is important to take account of both favourable and unfavourable situations. What would be the effect of the policy resolves in each of these contexts? After a preliminary confrontation, the policy resolves are adapted. This process is followed by new tests till the policy resolves have been sufficiently tested to be presented to Parliament or other policy-making bodies. In the course of testing, the people preparing policy have thus gone through a process of learning, the contexts serving as 'learning environments'.

Some developments in health care in Finland can be quoted as an example of such a learning process. At the end of 1981, in the framework of the 'Health for All by the Year 2000' programme of the World Health Organization, a steering group was installed in Finland. The steering group co-ordinates the activities of working groups of experts, each of which studies a certain aspect of health care. Variants are designed for each of the aspects. The working groups were given three background scenarios: a growth scenario, an adaptation scenario and a zero growth scenario. These alternative context scenarios deal with economic and social developments which are expected to exert an influence on the future of health care in Finland.

Murtomaa and Kanaanpää[1)] describe scenarios and their characteristics as follows:

> 'Scenarios are practical and relevant studies of the future. The characteristic feature of the scenario approach is that it sets out more than one possible future situation. The scenario provides insight into the general problems and those which are crucial for the future and crystallizes out the decisions which must considerably be made with regard to them.'

The working groups have explored more than thirty aspects of health care with the aid of the scenario method, using the year 2000 as time horizon. None of the scenarios designed however was directly concerned with the question of ageing. The bottle-necks which came to light when using scenarios in the preparation of policy related especially to bringing about social change. Is it for instance possible to systematically influence lifestyles and attitudes with respect to prevention?
Research as to the success or failure of campaigns aimed at influencing behaviour have made clear that results can be achieved provided certain conditions are met. For the sake of brevity we must refer readers to publications on 'planned change'.[2)]

The above may be summarized as follows. Policy-makers who desire to utilize scenarios on ageing as 'learning environments' for determining policy with respect to health care facilities for the elderly can find examples in the Finnish scenario projects. In these scenario projects the Finns make use of 'background scenarios' in which they describe autonomous developments. They investigate what improvements are possible and desirable in the areas of medicine and social medicine (cure and care). They also consider improvements in the area of prevention and lifestyles to be essential. The state of the art of conducting campaigns for influencing behaviour makes it possible - within certain limits - to consider lifestyles and other behavioural patterns as being subject to influence.

7.3 Scenarios and health care facilities for the elderly

Anyone who desires to use scenarios for screening alternative patterns of health care facilities for the elderly as to their desirability, would be well advised to start by listing the main bottle-necks with which the health of and health care facilities for the elderly in the Netherlands might be faced between 1984 and 2000. A list of problems which represents all the main points and succeeds in avoiding either exaggeration or underestimation, constitutes an indispensible point of departure for making an analysis. Especially the reference scenario in Chapter 2 makes clear where problems might be encountered.

A list of problems, even a short one for boosting the memory, should take account of at least the following elements.

- The state of health of the elderly, including the gap in life expectancy between men and women.
- The social position of the elderly.
- The level of facilities for the elderly.
- Collective expenditure for the elderly.

With this concisely formulated list of problems in mind, the scenarios will now be more closely examined as to their bearing on patterns of health care facilities for the elderly.
In this process, we can overlook those relationships already discussed in Chapters 2, 4 and 5. We now focus on possible influences in other cells of the scheme which has been included as Figure VII.1 (the scheme is also given in Chapter 1). The numbering of the cells shows the sequence in which they are discussed in this section.

Figure VII.1 Scenarios and patterns of health care facilities for the elderly

scenarios / patterns	scenario A (reference scenario)	scenario B (increasing growth of demand for facilities)	scenario C (decreasing (growth of demand for facilities)
1. maintaining present course	(see Chapter 2)	1	2
2. top care	3	(see Chapter 4)	4
3. towards a reciprocal aid society	5	6	(see Chapter 5)

To begin with, we take a look at cell 1. In how far is it possible to maintain the present course in the context of a growing demand for health care facilities for the elderly? What paths are open to the system of health care for coping with such developments? This scenario project is partly based on the realization that the present approach to health (care) policy is not adequate for dealing with the present autonomous developments. An increased demand for health care facilities would intensify the existing tensions. The consequences of this for the health care of the elderly would probably be a deterioration in this care. Long waiting lists for hospitals, nursing homes etc. would result. Only to a limited degree would the facilities be capable of absorbing the increased pressure. In such a situation, it would not be surprising if the demand for facilities seeks other outlets. It is not inconceivable that the market mechanism would assume a more important position with increased premiums for health care insurances etc. The elderly with a better income would more than is at present the case take recourse to commercial homes and possibly commercial hospitals. To counteract the negative social effects, it would be necessary to fall back on collective resources to a much greater extent.

We now take a look at cell 2. How would the pattern 'maintaining present course' be affected in the event of a decreasing demand for health care facilities? Adaptation of this pattern of facilities would be called for since the elderly would make less appeal to these facilities than in the past, while there would be a greater demand for another type of facilities (for instance, small scale facilities).[3] It is not unlikely that this would involve starting problems.

Cell 3 deals with the pattern of top care under conditions as outlined in the reference scenario. A combination of the reference scenario and top care as pattern of facilities will probably only be possible if the thresholds of facilities are raised by means of waiting lists, raising rates and premiums, to mention but a few measures which limit accessability. For the limited number of elderly persons who will still have access to facilities, there could be an improvement in the health situation. For all the rest who will then be excluded, a very unfavourable situation would arise. They would not be gradually stimulated to finding solutions for themselves, but be suddenly faced with the necessity for doing so. In addition, the unsatisfied demand might to some degree be met by commercial facilities.

Cell 4 sketches top care in combination with a declining demand for health care facilities. This could have the result that people only as a last resort turn to top care, after first having tried as far as possible to solve the problems themselves. By means of reciprocal aid, a change of mentality and change of lifestyles, the actual demand for health care facilities could be greatly reduced. The elderly who make use of top facilities would receive little stimulation to help themselves. In the shrinkage scenario, no serious problems would result from the demand for general resources for health care facilities.

In **cell 5** we have a combination of the present situation (reference scenario) with a pattern of health care facilities which inclines towards a reciprocal care society. The pattern of health care facilities would be comparatively new for the Netherlands. Starting problems would seem likely. Apart from these, this pattern of facilities might have a favourable effect on the health care of the elderly. Personal initiative on their part would be stimulated. Emancipation processes are part of this pattern. Especially in view of the relative novelty of the pattern, the initial demand made on collective resources would probably be quite great.

Cell 6 deals with an increasing demand as autonomous development and a pattern of facilities the ultimate goal of which is a reciprocal care society. It is very much to be questioned whether the growing demand would permit of a pattern of care with a strongly reciprocal character. This would be a combination which could only be realized with the greatest difficulty.

It would also be possible to incorporate the disturbing developments in the analysis of scenarios and patterns of health care facilities. Since in this section we only want to illustrate the possibilities of such an analysis, we will not undertake this expansion.

To what conclusions does the analysis lead? In the first place to reinforcing the analyses in Chapters 2, 4 and 5, in which the patterns 'maintaining present course', 'top care' and 'towards a reciprocal aid society' are linked to the reference scenario, the growth scenario and the shrinkage scenario respectively; in the second place to the conclusion that the context scenarios show a number of negative side effects in the case of most of the patterns of care examined. This calls for extra attention in policy making. The third conclusion is that during execution of health care policy, important changes in the situation may be expected. This makes it desirable to adhere to a flexible policy which offers possibilities for adaptation in the course of implementation. Since many processes have already been determined for the period between 1984 and 2000, there are limits to possibilities for adaptation. For instance the construction of a hospital is a job covering many years, which means that it is already for the most part known what new hospitals will come available in the nineties.

7.4 A closer look at putting the scenario report to active use

Naturally, the scenario report could lend support to those people responsible for drawing up the **Policy Memorandum on Health in 2000**. They will be able to view their analyses and policy designs against the background of the scenarios presented. As soon as the Policy Memorandum is published, readers of this memorandum too may benefit from the scenario report. On the basis of the report, it would be possible for them to estimate to what degree the Policy Memorandum goes along with autonomous developments which may be expected with respect to the elderly and their health. After the Policy Memorandum has been dealt with in Parliament, a new phase opens up. By this time implementation of policy should be in progress.

If users should desire to devote intensive thought to the size and direction of investments in medical and social scientific research relating to the elderly, it would be necessary to supplement both the background study and the scenarios. This means that one would have to devote attention not only to patterns of health care facilities for the elderly, but also to scientific activities, especially in connection with the realization of the patterns involved.

Supplementing the background study and the scenarios could also mean that processes come under attention which are the consequence of private enterprise increasingly taking over health care for the elderly. A decrease of the effective demand for health care facilities might for instance be brought about not only by a change of mentality, but by raising admission thresholds (commercial hospitals, etc.).

The scenarios in this report might also be put to good use in the designing and implementing of an early warning system. This would in the first place be useful in connection with the desirability to recognize new medical and medical-technological developments as quickly as possible, and, thanks to an anticipatory policy, to be ready to react to them; and secondly to recognize new developments on the social level, for instance new forms of intentional influencing of lifestyles, as well as new forms of activities and patterns of co-operation in society. There should for instance be timely recognition of a greater desire than formerly for the elderly to participate in the labour process. Once this desire is recognized, it can be further investigated what stimulating policy activities are advisable.

7.5 Concluding remarks

As was stated in Chapter 1, the first research question of the scenario project is:

> "What are likely to be the most important (future) developments which will exert an influence on the health situation of the elderly in the Netherlands in the period 1984-2000?"

The project has made clear that Dutch health (care) policy with respect to the elderly will have to be prepared for autonomous developments of a strongly diverging nature. Only in a few areas - especially demographic developments - is it possible to make forecasts which are sufficiently reliable and precise to serve as an unambiguous basis for policy. The expected autonomous developments call for a strategy which can survive reversals (preparing for the worst), and which leaves scope for the necessary adjustments in the course of implementation till the year 2000. Such a flexible policy would be most effective if combined with a system of monitoring - following developments with the aid of social indicators, etc. - and with an 'early warning system' - the early recognition of reversals or new trends in the autonomous developments.

When determining the aspects to be stimulated, it is advisable not to stimulate only the strictly medical and medical-technological innovations which promote care of the elderly. Possibilities for altering social factors such as lifestyles also call for systematic stimulation.

We now take a look at the second research question from which the scenario project proceeded:

> "In view of the future health situation of the elderly and their increasing share in the Dutch population, what are the possible patterns of (health) care facilities in the period 1984-2000?"

The results of this project do not provide an unambiguous recommendation with respect to the choice of a pattern of health care facilities for the elderly. The second research question led to an answer which calls for further choices. The scenarios provide clues for making these choices and clarify the consequences of alternatives.

Notes Chapter 7

(1) Murtomaa and Kankaanpää, 1984. In drawing up the Dutch health (care) scenarios, Finnish experience was also utilized, see Pannenborg, 1984; scenario committee on coronary and arterial diseases, 1984; scenario committee on cancer, 1984; scenario committee on lifestyles, 1984.

(2) Boruch and Riecken, 1975; Fairweather and Tornatzky, 1977; Van Vught, 1982.

(3) See also for instance Lägergren, 1984.

Appendices

APPENDIX A

The Dutch population; share of persons older than 55 years of age in the total population, differentiated according to age.
The age groups are given as percentages of the total number of persons older than 55. The column older than 55 is given as a percentage of the total population. Absolute number x 1000.

year	total population	index	older than 55	55-59	60-64	65-69	70-74	75-79	80-84	`85
1983	14385	100	3039	707	640	545	457	341	209	140
			21.1	23.3	21.1	17.9	15.0	11.2	6.9	4.6
1984	14473	101	3078	704	664	538	463	348	215	146
			21.3	22.9	21.6	17.5	15.0	11.3	7.0	4.7
1985	14568	101	3114	704	681	533	469	354	222	151
			21.4	22.6	21.9	17.1	15.1	11.4	7.1	4.8
1986	14646	102	3158	713	676	555	471	358	229	156
			21.6	22.6	21.4	17.6	14.9	11.3	7.3	4.9
1987	14730	102	3193	718	672	572	475	360	234	162
			21.7	22.5	21.0	17.9	14.9	11.3	7.3	5.1
1990	14973	104	3292	729	668	626	464	376	248	181
			22.0	22.2	20.3	19.0	14.1	11.4	7.5	5.5
2000	15643	109	3663	857	717	632	534	437	261	225
			23.4	23.4	19.6	17.3	14.6	11.9	7.1	6.1
2010	15801	110	4415	1065	1046	743	570	439	299	253
			27.9	24.7	24.2	17.2	13.2	10.2	4.6	5.9
2020	15697	109	5077	1178	1053	921	830	514	317	264
			32.3	23.2	20.8	18.2	16.3	10.1	6.2	5.2
2030	15350	107	5365	1025	1092	1016	835	636	462	299
			35.0	19.1	20.3	18.9	15.6	11.9	8.6	5.6
INDEX:			177	145	171	186	183	187	221	214

The age columns 55-59 through `85 fall under the heading AGE CATEGORIES.

For purposes of indexing, the year 1983 is set at 100.

Source: Assembled from the 'Forecast of Population Development after 1980'. CBS, The Hague, 1982. Figures refer to the medium variant.

APPENDIX B

Population of 55 and older according to civil status of 1st January 1980, 1st January 2000 and 1st January 2030 (according to medium variant CBS population forecast) x 1000

1-1-1980

Age	Unmarried		Married		Widowed		Divorced		Total	
	M	F	M	F	M	F	M	F	M	F
55-59	22	29	303	285	9	46	11	15	345	375
60-64	16	27	237	209	12	62	8	12	273	309
65-69	15	30	203	170	17	88	6	11	242	298
70-74	12	29	147	114	23	106	4	8	186	258
75-79	9	23	88	61	26	106	2	5	125	194
80-84	4	14	40	24	24	80	1	3	70	121
85 +	3	10	16	7	24	59	1	2	44	77
Total	81	162	1034	870	135	547	33	56	1285	1632

1-1-2000

Age	Unmarried		Married		Widowed		Divorced		Total	
	M	F	M	F	M	F	M	F	M	F
55-59	37	21	347	310	12	45	39	46	435	422
60-64	28	19	283	244	16	65	28	35	354	363
65-69	22	20	234	196	22	93	19	27	296	336
70-74	15	22	175	141	28	121	12	20	230	304
75-79	10	21	118	90	34	142	7	15	169	268
80-84	5	15	52	35	30	113	3	8	90	171
85 +	4	17	23	14	35	123	2	6	65	160
Total	121	135	1232	1030	177	702	110	157	1639	2024

1-1-2030

Age	Unmarried		Married		Widowed		Divorced		Total	
	M	F	M	F	M	F	M	F	M	F
55-59	144	92	314	316	11	46	42	60	511	514
60-64	146	101	325	311	19	81	43	67	533	559
65-69	122	94	289	260	27	120	37	66	476	540
70-74	79	68	223	185	36	155	28	59	367	468
75-79	41	38	152	114	45	177	19	49	257	379
80-84	18	19	88	58	51	179	12	37	169	293
85 +	7	11	32	18	49	154	6	22	94	205
Total	557	423	1423	1262	238	912	187	360	2407	2958

APPENDIX C

Independently residing elderly persons who suffer from various ailments according to nature of the ailment, age category and sex.

	rheuma-tism	chronic cough	short-ness of breath	heart com-plaints	high BP	conse-quences of stroke	dizzi-ness	dia-betes	anae-mia	conse-quences of accident	asthma	total (=100%)
men												
55-59	24	11	18	13	15	2	11	3	1	6	3	382
60-64	27	12	24	15	15	3	8	8	3	6	5	353
65-69	26	14	24	15	15	2	11	6	1	6	4	342
70-74	22	14	25	17	10	4	10	6	1	9	6	332
75-79	25	17	30	18	11	4	15	5	0	5	2	269
80 and older	30	19	35	19	16	3	26	6	4	6	6	193
total 55 a.o.	25	14	24	15	14	3	12	6	1	6	4	1871
total 65 a.o.	25	15	27	17	13	3	14	6	1	6	5	1136
women												
55-59	34	6	12	5	20	1	13	4	6	4	2	388
60-64	36	7	13	9	20	1	13	4	4	6	3	473
65-69	36	9	19	8	26	3	20	7	4	7	3	426
70-74	42	9	25	15	30	2	21	9	3	7	5	419
75-79	48	9	25	15	35	3	23	11	4	4	3	423
80 and older	45	9	25	15	20	2	29	8	5	7	2	283
total 55 a.o.	39	8	18	10	25	2	19	6	4	6	3	2412
total 65 a.o.	42	9	23	13	290	3	23	9	4	6	3	1551
men and women												
total 55 a.o.	33	10	21	12	20	2	15	6	3	6	3	4283
total 65 a.o.	35	12	25	14	22	3	19	7	3	6	4	2687

Source: CBS: The Living Conditions of the Population of 55 years of Age and Older - 1982.

Source: CBS: The Living Conditions of the Population of 55 years of Age and Older - 1982.

APPENDIX D: Mortality according to cause, 1983

The data which follow have been derived from the CBS and relate to the year 1983. (CBS, Deaths according to cause, age and sex, series B1, AM list (adapted mortality list).
Data are given with respect to:

Table D-1: absolute number of deaths according to cause (AM list), age and sex.

Table D-2: share of the population of 55 and older in the total number of deaths resulting from a certain cause (abridged), according to sex (data processed by us).

Table D-3: share of a certain cause of death (abridged) in total mortality of the elderly ($>$ 55 years of age), according to sex (data processed by us).

Table D-1: absolute number of deaths according to cause (AM list), age and sex

cause of death	sex	tot.	<55	55-59	60-64	65-69	70-74	75-79	80-84	85+
1 Infection diseases of alimentary tract	m	17	3	1	1	1	1	2	4	4
	f	32	6		1	2	2	3	6	12
2 Tuberculosis	m	28	3		2	6	2	3	8	4
	f	12	2		1	3		1	2	3
3 Latent effects of tuberculosis	m	78	3	4	7	13	8	17	14	12
	f	43	2	4	3	7	8	7	5	7
4 Whooping cough	m	0								
	f	0								
5 Mengingococcus infections	m	4	3					1		
	f	3	1	1			1			
6 Tetanus	m	0								
	f	1		1						
7 Septicemia	m	94	12	4	8	10	12	22	15	11
	f	144	13	5	10	8	26	23	25	34
8 Smallpox	m	0								
	f	0								
9 Measles	m	0								
	f	1	1							
10 Malaria	m	1		1						
	f	0								
11 Other infectuous and parasitic diseases	m	77	24	6	9	8	7	6	6	11
	f	103	20	5	2	5	8	13	13	37

Table D-1: absolute number of deaths according to cause (AM list),
(contin.) age and sex

cause of death	sex	tot.	<55	55-59	60-64	65-69	70-74	75-79	80-84	85+
12 Malignant neoplasms	m	1561	128	90	169	203	288	295	216	172
(stomach)	f	946	66	43	49	104	132	183	165	204
13 Malignant neoplasms	m	1246	106	83	127	167	197	213	197	156
(colon)	f	1553	128	92	118	165	208	287	269	286
14 Malignant neoplasms	m	572	39	37	45	63	105	112	89	82
(rectum)	f	449	38	17	45	39	81	63	82	84
15 Malignant neoplasms	m	7104	536	603	967	1292	1507	1201	655	343
(trachea, lungs)	f	800	163	79	103	83	106	110	80	76
16 Malignant neoplasms (mamma)	f	2897	648	337	339	344	306	317	304	302
17 Malignant neoplasms (cervix uteri)	f	300	76	30	39	50	34	31	22	18
18 Leukemia	m	522	131	37	42	67	60	68	64	53
	f	452	97	22	40	36	44	62	76	75
19 Other neoplasms	m	7728	972	547	799	1021	1278	1272	934	905
	f	5984	731	431	564	691	892	990	867	818
20 Diabetes mellitus	m	474	63	35	41	50	53	61	80	91
	f	873	41	35	54	80	98	140	185	240
21 Nutritional	m	18	1	1	1		3	5		7
marasmus	f	33	3	1	1	1	3	5	4	15
22 Other protein	m	1					1			
malnutrition	f	3					1			2

Table D-1: absolute number of deaths according to cause (AM list),
(contin.) age and sex

cause of death	sex	tot.	<55	55-59	60-64	65-69	70-74	75-79	80-84	85+
23 Anemias	m	78	5	2	2	11	11	14	10	23
	f	97	9	1	3	7	5	16	19	37
24 Meningitis	m	26	17		1	2	1	2	2	1
	f	35	19		1	1	5	4	3	2
25 Acute arthritis	m	0								
	f	0								
26 Chronic rheumatic heart disease	m	71	14	4	11	16	11	11	3	1
	f	119	16	15	13	23	26	15	8	3
27 Hypertension	m	313	36	36	29	43	49	52	38	30
	f	420	15	8	27	33	58	76	94	109
28 Acute myocardial infarct	m	12143	1296	979	1419	1833	2056	1988	1377	1195
	f	7726	243	206	387	705	1181	1624	1694	1686
29 Other ischemic heart complaints	m	2805	170	187	302	365	485	535	405	356
	f	1931	44	38	83	127	228	374	472	565
30 Cerebrovascular complaints	m	5059	243	147	260	479	795	1055	1035	1045
	f	6722	215	112	184	357	725	1245	1658	2226
31 Arteriosclerosis	m	414	4	3	17	34	45	71	86	154
	f	439	3	4	6	13	17	61	88	247
32 Other circulatory disorders	m	6971	418	269	437	664	1003	1253	1225	1702
	f	7283	150	106	184	309	653	1189	1683	3009

Table D-1: absolute number of deaths according to cause (AM list),
(contin.) age and sex

cause of death	sex	tot.	<55	55-59	60-64	65-69	70-74	75-79	80-84	85+
33 Pneumonia	m	1353	37	21	21	59	124	220	309	562
	f	1647	29	8	16	25	77	164	371	957
34 Influenza	m	89	7	2	2	5	6	10	18	39
	f	99	5	1		1	5	14	19	54
35 Bronchitis/ emphysema/asthma	m	2152	69	73	149	266	372	515	365	343
	f	607	38	30	34	49	65	115	100	176
36 Peptic ulcer	m	227	20	9	17	29	34	49	33	36
	f	243	8	4	8	10	26	43	64	80
37 Appendicitis	m	24	4	1	1	4	1	4	6	3
	f	24	4		1	1	4	5	4	5
38 Cirrhosis of the liver	m	475	163	56	56	67	52	38	29	14
	f	280	72	31	28	31	31	34	39	14
39 Nephritis/ nephrosis	m	609	14	9	15	48	52	97	124	250
	f	775	12	6	15	38	63	115	151	375
40 Prostate hyper-trophy	m	247		1	5	7	36	42	61	95
41 Abortion	f	1	1							
42 Obstetrical causes	f	7	7							
43 Congenital defects	m	421	391	6	9	1	3	7	2	2
	f	349	295	13	12	8	7	5	5	4

Table D-1: absolute number of deaths according to cause (AM list), (contin.) age and sex

cause of death	sex	tot.	<55	55-59	60-64	65-69	70-74	75-79	80-84	85+
44 Birth traumas	m	36	36							
	f	19	19							
45 Other perinatal causes	m	288	288							
	f	214	214							
46 Symptoms/ill-defined diagnosis	m	2204	630	162	184	202	256	210	195	365
	f	1817	361	58	81	106	140	181	239	651
47 Other diseases	m	5139	585	190	290	460	713	907	918	1076
	f	5537	435	136	231	331	533	843	1119	1909
48 Motor accidents on public roads	m	1211	857	58	45	55	64	64	50	18
	f	478	261	28	36	34	55	27	30	7
49 Other motor accidents	m	17	14	1		1				1
	f	1	1							
50 Accidental falls	m	551	65	14	20	24	42	69	107	210
	f	950	19	4	11	19	38	118	220	521
51 Suicide/auto-mutilation	m	1039	625	88	85	56	63	52	39	31
	f	681	369	62	61	50	58	39	23	19
52 Murder/mutilation	m	86	79	1	1	1	1	1	1	1
	f	49	37	1	2	3	1	2	3	
53 Other external causes	m	696	476	33	37	28	35	32	28	27
	f	313	163	12	14	12	19	27	23	43
Total	m	64269	8587	3801	5633	7661	9832	10576	8748	9431
	f	53492	5100	1987	2807	3911	5970	8571	10234	14912

Table D-2: share of the population of 55 and older in the total number of deaths resulting from a certain cause (abridged), according to sex

		absolute number of deaths among the elderly		deaths of the elderly as a percentage of total mortality	
nrs.	cause of death	male	female	male	female
1-11	Infections and parasitic diseases	251	294	83.9	86.7
12-19	Malignant neoplasms	16821	11434	89.8	85.4
20-22	Endocrine, nutritional and metabolic diseases	429	865	87.0	95.2
23	Anaemia	73	88	93.6	90.7
24	Meningitis	9	16	34.6	45.7
26-32	Diseases of the circulatory system	25595	23954	92.1	97.2
33-35	Diseases of the respiratory system	3481	2281	96.9	96.9
36-38	Diseases of the digestive system	539	463	74.2	84.6
39-40	Diseases of the genitourinary system	842	763	98.4	98.5
41-45	Abortion, other obstetric causes, congenital defects, birth traumas and other perinatal defects	30	4	4.0	9.2
46	Symptoms/ill-defined diagnoses	1574	1456	71.4	80.1
47	Other diseases	4554	5102	88.6	92.1
48-53	External causes of death	1484	1622	41.2	65.6
Total mortality of elderly (≥ 55 years)		55682	48392	86.6	90.5

Table D-3: share of a certain cause of death (abridged) in total deaths of the elderly (⩾ 55), according to sex

cause of death		absolute		percentag	
		male	female	male	fe
nrs. 1-11	Infectious and parasitic diseases	251	294	0.5	
nrs. 12-19	Malignant neoplasms	16821	11434	30.2	
nrs. 20-22	Endocrine, nutritional and metabolic diseases	429	865	0.8	
nr. 23	Anaemia	73	88	0.1	
nr. 24	Meningitis	9	16		
nrs. 26-32	Diseases of the circulatory system	25595	23954	46.0	
nrs. 33-35	Diseases of the respiratory system	3481	2281	6.3	
nrs. 36-38	Diseases of the digestive system	539	463	1.0	
nrs. 39-40	Diseases of the genitourinary system	842	763	1.5	
nrs. 41-45	Abortion, other obstetric causes, congenital defects, birth traumas and other perinatal defects	30	54	0.1	
nr. 46	Symptoms/ill-defined diagnoses	1574	1456	2.8	
nr. 47	Other diseases	4554	5102	8.2	
nrs. 48-53	External causes of death	1484	1622	2.7	
Total mortality of elderly (⩾ 55 years)		55682	48392	100.0	1(

APPENDIX E: OVERVIEW OF DISEASES AMONG THE ELDERLY

The overview given below is intended to provide some insight into diseases which play an important role among the elderly. We discuss diseases which are numerically important as well as those peculiar to the elderly. The international statistical classification of diseases, injuries and causes of death (ICD, 1975)[1] has been employed. In as far as these are given in literature on the elderly, prevalence figures are presented. No systematic analysis of epidemiological literature was carried out for this purpose. Appendix D (deaths according to cause) and Appendix H (discharge diagnoses of elderly hospital patients) may also provide some insight into the morbidity of the elderly.

II **Neoplasms** (especially malignant)

- Most frequent localities[2] among men over 65 years of age: bronchi, alimentary tract, prostate (men of 70: 8%[3]), skin and bladder.
- Most frequent localities among women over 65 years of age: mamma, uterus and ovaries, digestive tract and skin.

III **Endocrine, nutritional and metabolic diseases and immunity disorders**

- autoimmune diseases
- variant of diabetes mellitus encountered among the elderly
  - men (70 years old): 5%)
  - women (70 years old): 6%)[3]
  - men and women (` 65 years): 13%-50%[4]
  - data from GP practices:[5]

number/1000/patients/year

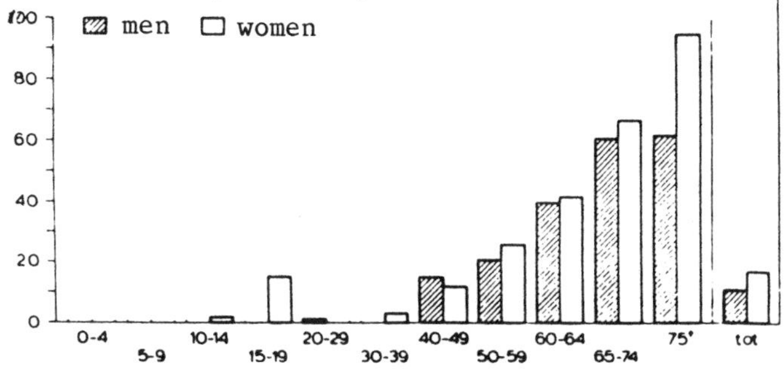

men women

100 80 60 40 20 0

0-4 5-9 10-14 15-19 20-29 30-39 40-49 50-59 60-64 65-74 75+ tot

age categories

- hypothyroidism

V Mental disorders

- depression: varying from 2 or 3% to 65%[6)]
- dementia:
  . Senile Dementia of the Alzheimer Type (SDAT)
  . vascular dementia (also known as multi-infarct or arterio-sclerotic dementia).

  (for figures see Chapter 6, Section 6.2.2).

VI Diseases of the nervous system and sense organs

- Parkinson's disease
  . data from GP practices[7)]:

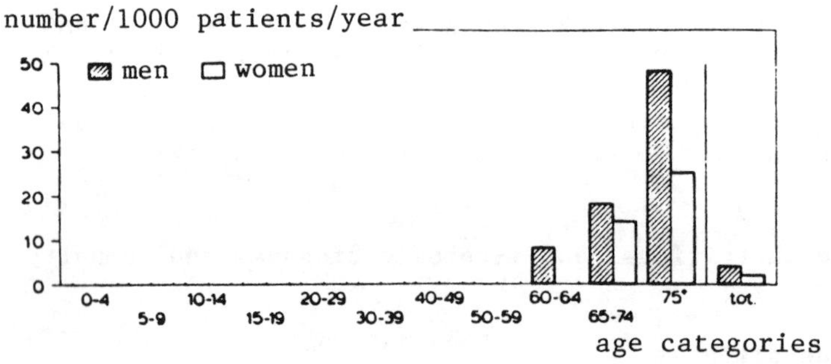

- defects of hearing
  . data from GP practices[7)]:

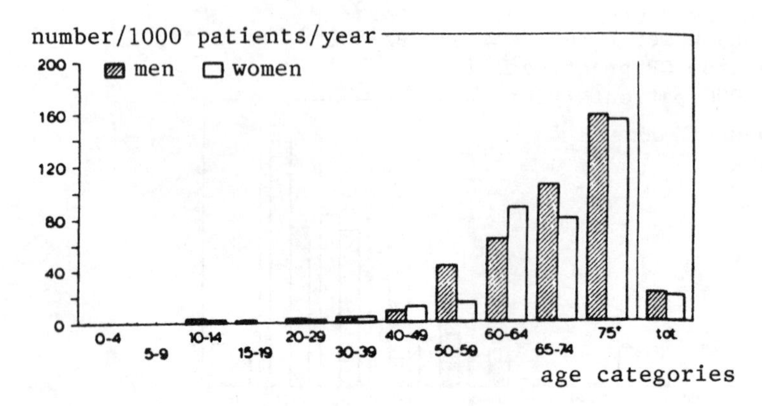

- 1/3 to 1/2 of the elderly[8)]
- independently residing persons over 55: 19%)
  residents of homes for the elderly over 55: 42%)[9)]
  men more defects than women)

- defects of vision:
  - women more defects than men)
  - nearby)
    independently residing persons over 55 15%)
    residents of homes for the elderly over 55 49%)[10)]
  - at a distance)
    independently residing persons over 55 13%)
    residents of homes for the elderly over 55 41%
  - nearby: 20% of the elderly[8)]
  - glaucoma simplex: 2-7%[11)]
  - senile macula degeneration
  - aneurysms
  - defects of the retina
  - cataract: data from GP practices[12)]:

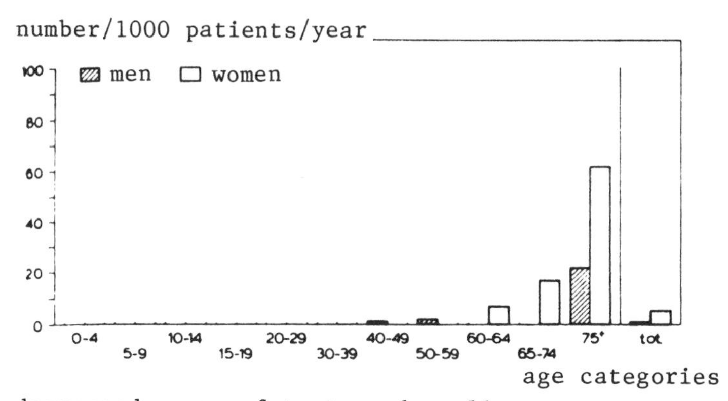

- decreased sense of taste and smell

VII Diseases of the circulatory system

- hypertension[13)]
  - men varying from 8.1% in the age category)
    60-64 to 0.7% in the age category 85-89)
  - women varying from 10.3% among the 'younger')[14)]
    elderly to 12.3% among the 'older' ones)
  - data from GP practices[15)]:

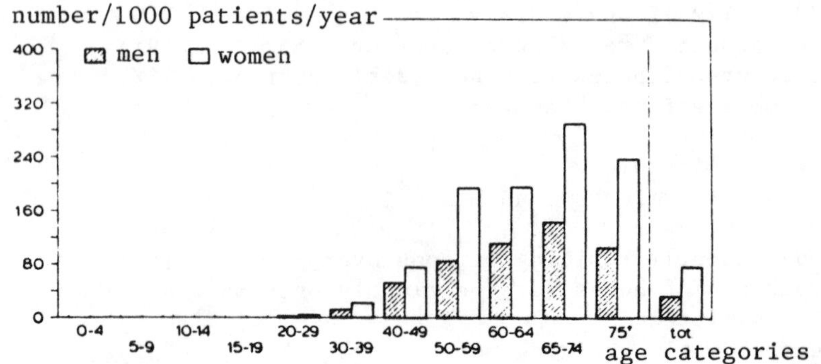

- ischemic heart diseases: over 60 years of age: 10-15%[14])
- cerebrovascular disorders:
  - data from GP practices[16])

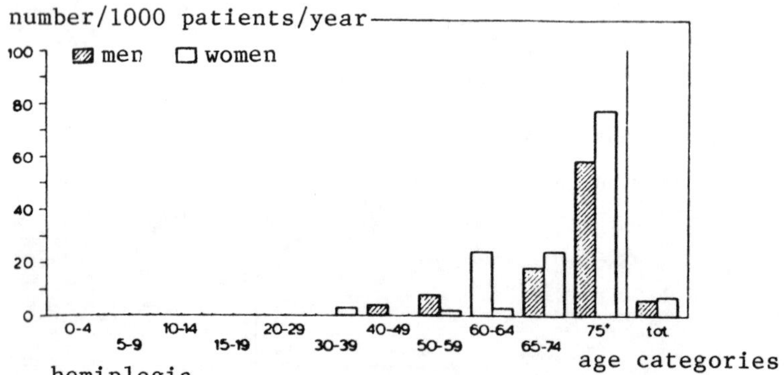

  - hemiplegia
  - hemiparesis: men (70 years of age) : 1%)
women (70 years of age): 1%)[3])

- arteriosclerosis
- giant cell arteritis
- résumé of data[17])

age group	any cardio-vascular %	arterial hyper-tension %	ischemic heart disease %	cardiac failure %	cerebro-vascular disease %
65-69	43.7	21.7	12.5	16.3	5.0
70-74	48.3	22.8	12.2	23.6	4.8
75-79	59.4	20.8	11.9	38.6	5.6
80-84	54.2	18.2	8.9	37.5	7.8
85-89	60.3	6.9	12.1	37.9	5.2
90-99	66.7	6.7	13.3	60.0	6.7

VIII Diseases of the respiratory system

- chronic bronchitis
  - men (of 70) 18%)
  - women (of 70) 9%)[3]
  - data from GP practices[18])

number/1000 patients/year

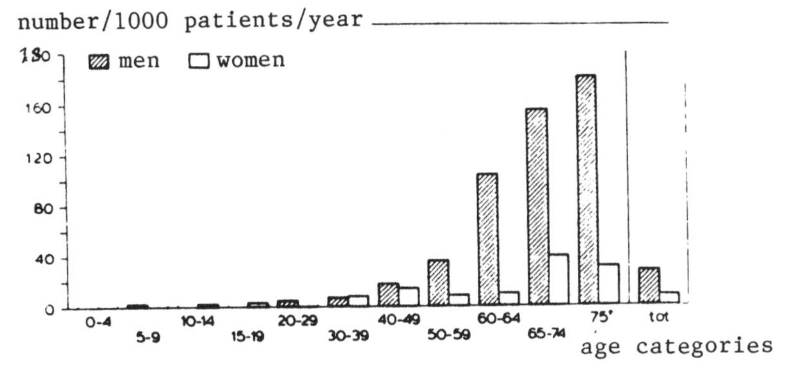

age categories

- résumé of data[17])

age group	any respiratory disease %	chronic bronchitis %	pulmonary emphysema %	pulmonary tuberculosis %
65-69	9.8	1.9	3.2	0.9
70-74	9.4	2.8	2.0	0.6
75-79	6.4	1.1	1.4	0.6
80-84	6.8	1.0	2.6	0.5
85-89	1.7	0.0	0.0	0.0
90-99	6.7	6.7	0.0	0.0

X Diseases of the genitourinary system

- prostate hypertrophy
  - men (of 70): 40%[3])
  - data from GP practice[19]):

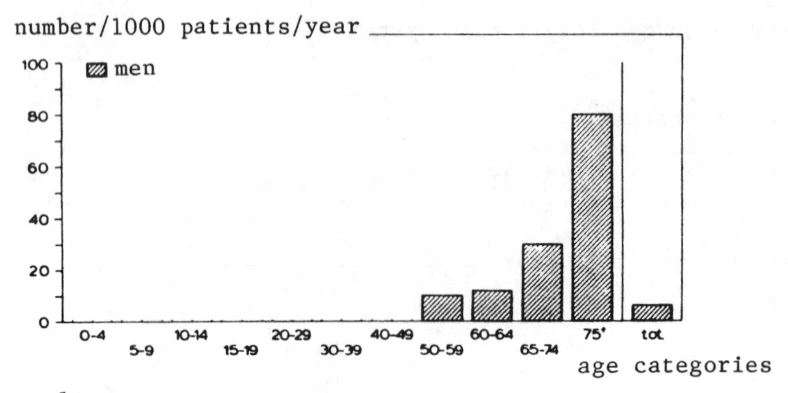

- prolapse
  . data from GP practice[19]:

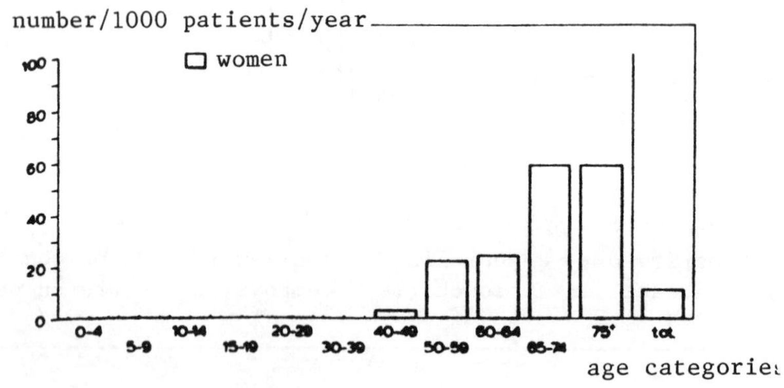

- kidney insufficiencies:
- incontinence of urine
  . on laughing/coughing : women of 70: 44%)
    men of 70 : 6%)
  . at night : women of 70: 1%)[3]
    men of 70 : 2%
  . poor control over musculus sphincter:
    women over 60 : 8-27%)
    men over 60 : 5-22%)[8]

XII Diseases of the skin and subcutaneous tissue

- senile atrophy of the skin (shrivelling)
- decubitus

XIII Diseases of the musculoskeletal system and connective tissue

- rheumatoid arthritis
  . women (of 70): 6%)
  . men (of 70): 3%)[3]
  . data from GP practice[20]):

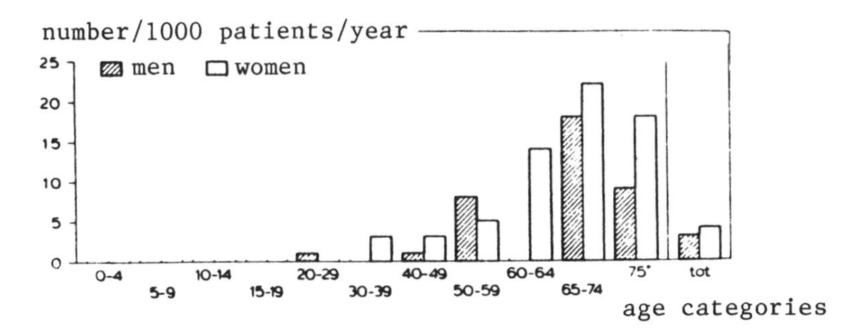

- arthrosis deformans
  . data from GP practice[21]):

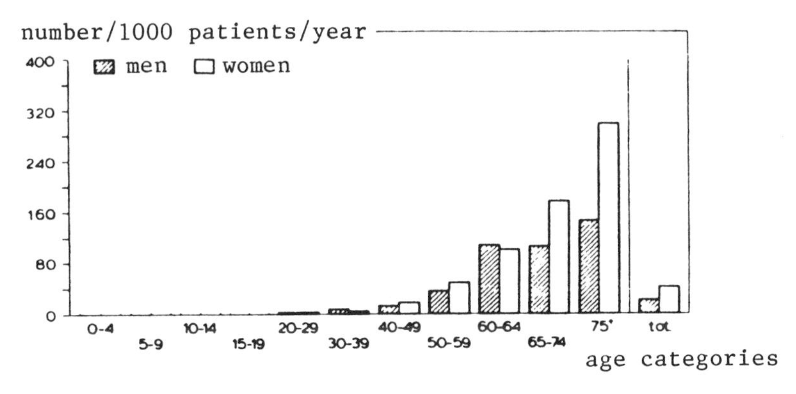

- osteoporosis (especially among women):

  . percentage of patients of 65 and older admitted to hospitals with osteoporosis[22]):

	men	women
1972	0.39	1.19
1977	0.51	1.40
1982	0.69	1.70

- muscular atrophy

XVI Symptoms, signs and ill-defined conditions. External causes of injury and poisoning

- fractures (of the hip) (inter alia resulting from osteoporosis)
- drug poisoning
- (traffic) accidents

Lastly we show the dental conditions encountered among the elderly. The ICD classification is not employed here.

Dental problems

- difficulty with chewing over the age of 60: 38.5%[23)]
- dental prostheses[24)]

	men	women
55-59	83.5%	82.9%
60-64	81.8%	99.0%
65-69	91.0%	92.6%
70-74	90.1%	95.6%
75-79	89.0%	91.6%
80 +	81.8%	90.6%

- problems with biting (a hard apple for instance) among people with dentures[24)]

	men	women
60-69	50.0%	49.8%
70 +	66.0%	61.5%

NOTES APPENDIX E

(1)
I Infectious and parasitic diseases
II Neoplasms
III Endocrine, nutritional and metabolic diseases and immunity disorders
IV Diseases of the blood and blood-forming organs
V Mental disorders
VI Diseases of the nervous system and sense organs
VII Diseases of the circulatory system
VIII Diseases of the respiratory system
IX Diseases of the digestive system
X Diseases of the genitourinary system
XI Complications of pregnancy, childbirth and puerperium
XII Diseases of the skin and subcutaneous tissue
XIII Diseases of the musculoskeletal system
XIV Congenital anomalies
XV Certain conditions originating in the perinatal period
XVI Symptoms, signs and ill-defined conditions
External causes of injury and poisoning.

(2) Sillevis Smit, 1975, p. 31.

(3) Data of the Gothenburg survey among 70-year-olds carried out by Svanborg. Quoted by WHO, 1982a, pp. 14-15.

(4) Quoted by Fuldauer, 1973, p. 143.

(5) Voorn, 1983, p. 84. These are prevalence data dating from 1978, deriving from the practices of four GPs affiliated to the continuous morbidity registration of the Nijmegen Inter-university Institute of General Practitioners. The four practices (N=12,000 patients) shoud be reasonably representative of the Dutch population as regards distribution according to sex and age. Voorn uses the concept 'period pre-valence': the number of (known and new) cases of disease in a certain population during a certain period (namely one year).

(6) Quoted by Godderis, 1983, p. 306. Variations in data with respect to depression are connected with the origin of the group examined as well as the severity of the symptomatology and the duration of the periods of depression which are used as criteria.

(7) Voorn, 1983, pp. 132 and 138.

(8) Derived from a WHO survey in eleven countries, quoted in WHO, 1983, p. 53.

(9) CBS, 1984, pp. 96-97. The percentages quoted here are a summation of the categories moderate and bad.

(10) CBS, 1984, pp. 94-96. The percentages quoted here are a summation of the categories moderate, bad, and not at all.

(11) Fuldauer, 1973, p. 143.

(12) Voorn, 1983, p. 135.

(13) There is a lack of consistency in data on hypertension partly due to the fact that there are no adequate diagnostic criteria, especially with respect to the elderly.

(14) Derived from a survey in Yugoslavia, quoted in WHO, 1982b, p. 6.

(15) Voorn, 1983, p. 148.

(16) Voorn, 1983, p. 123.

(17) Quoted in WHO, 1982b, pp. 16 and 17. Derived from Finland.

(18) Voorn, 1983, p. 184.

(19) Voorn, 1983, pp. 203 and 213.

(20) Voorn, 1983, p. 242.

(21) Voorn, 1983, p. 235.

(22) According to data of the Foundation for Medical Registration (SMR) and the CBS.

(23) Survey in Jerusalem, quoted in WHO, 1983, p. 53.

(24) CBS, 1982/10, pp. 13 and 14. That there is a decrease in the number of persons with a dental prosthesis in the highest age categories can probably be explained by the fact that a number of persons in this age category, though having lost their natural teeth, have not, or no longer have, a prosthesis.

APPENDIX F

Independently residing persons according to ability to perform ADL, age category and sex - 1976

	all activ. without diff.	one activ. with diff.	2 or more activ. with diff.	can't wash complete-ly	can't dress or un-dress	can't eat or drink alone	can't perform 1 or more oth.activ.	unknown	total	
men										abs.
55-59	91.0	3.5	3.7	0.8	0.6	-	0.2	0.2	100	509
60-64	91.9	4.2	1.9	0.9	0.5	0.2	0.2	0.2	100	430
65-69	88.7	4.3	4.5	0.9	0.7	-	0.7	0.2	100	443
70-74	86.1	4.0	5.0	1.3	1.8	0.3	0.8	0.8	100	397
75-79	82.7	6.6	4.6	1.7	1.4	0.3	0.3	2.3	100	340
80 and older	73.9	6.4	4.7	6.4	4.1	1.0	1.0	2.4	100	295
total 55 and older	88.1	4.4	3.8	1.4	1.1	0.2	0.4	0.6	100	2420
total 65 and older	84.9	4.9	4.7	1.9	1.6	0.3	0.7	1.0	100	1481
women										
55-59	91.3	2.2	3.9	0.4	1.0	-	0.6	0.6	100	508
60-64	89.3	3.5	4.6	0.4	1.1	0.2	0.7	0.2	100	541
65-69	84.6	4.9	5.7	1.3	1.1	0.2	0.7	1.5	100	546
70-74	78.1	7.5	7.2	1.5	2.5	0.2	2.3	0.8	100	530
75-79	73.0	6.7	7.7	3.3	1.6	0.2	5.1	2.4	100	508
80 and older	52.7	10.2	9.9	12.6	4.4	0.7	7.1	2.4	100	294
total 55 and older	83.2	4.9	5.7	1.8	1.5	0.2	1.8	1.0	100	2927
total 65 and older	76.9	6.7	7.0	3.0	2.0	0.2	2.8	1.5	100	1878

Residents of homes for the elderly according to ability to perform ADL, age category and sex - 1976

	all activ. without diff.	one activ. with diff.	2 or more activ. with diff.	can't wash complete-ly	can't dress or un-dress	can't eat or drink alone	can't perform 1 or more oth.activ.	unknown	total	
men										abs.
55-74	62.1	12.2	3.8	9.1	9.1	3.8	-	-	100	23
75-79	68.3	2.4	4.9	17.1	4.9	-	2.4	-	100	41
80 and older	42.6	2.1	5.3	34.0	13.8	-	2.1	-	100	94
total 55 and older	53.4	4.5	4.8	24.1	10.6	0.9	1.7	-	100	158
women										
55-74	62.6	6.6	5.8	10.3	11.4	-	1.7	1.7	100	56
75-79	48.2	8.2	8.2	21.2	8.2	1.2	4.7	-	100	85
80 and older	32.0	5.5	10.5	33.5	8.0	1.5	7.0	2.0	100	200
total 55 and older	43.3	6.4	8.8	24.9	8.9	1.1	5.2	1.4	100	341

Source: CBS, 1977, p. 55

APPENDIX F (continued)

Independently residing persons according to ability to perform ADL, age category and sex - 1982

	all activ. without diff. and without aid	one activ. only with diff. and all without aid	two or more activ. only with diff. and all without aid	one activ. only with aid	two or more activ. only with aid	unknown	total (=100%)
men							
55-59	86	4	7	2	1	1	382
60-64	86	7	5	1	1	1	353
65-69	83	7	6	1	2	0	342
70-74	77	9	7	2	3	2	332
75-79	74	10	10	2	3	1	269
80 and older	62	13	12	5	7	1	192
total 55 and older	81	7	7	2	2	1	1871
total 65 and older	77	9	8	2	3	1	1136
women							
55-59	86	7	6	1	0	0	388
60-64	82	8	7	3	1	0	473
65-69	80	7	9	1	3	1	426
70-74	71	11	14	2	3	0	419
75-79	56	16	16	6	5	0	423
80 and older	47	15	19	9	9	1	283
total 55 and older	74	10	11	3	3	9	2412
total 65 and older	66	11	14	4	4	0	1551
men and women							
total 55 and older	77	9	9	2	2	1	4283
total 65 and older	71	10	11	3	4	1	2687

Residents of homes for the elderly according to ability to perform ADL, age category and sex - 1982

	all activ. without diff. and without aid	one activ. only with diff. and all without aid	two or more activ. only with diff. and all without aid	one activ. only with aid	two or more activ. only with aid	unknown	total (=100%)
men							
55-79	44	11	2	22	21	0	33
80 and older	28	17	14	12	25	4	82
total 55 and older	34	15	10	16	23	2	115
women							
55-79	32	18	13	9	24	3	78
80 and older	13	9	8	23	47	1	155
total 55 and older	19	12	10	19	39	2	233
men and women							
total 55 and older	23	13	10	18	35	2	348

Source: CBS: Survey 'The Living Conditions of the Population of 55 and Older' - 1982.

APPENDIX G

The 20 most often presenting complaints among men in a 'standard GP practice' (n=2800) in the years 1980 and 2000. Concerns **all** complaints among **men**.

Rank 2000	Rank 1980	abs. nrs. 2000	abs. nrs. 1980	Presenting complaints	percentage incr.	percentage decr.
1	1	281	275	Non-febrile comm.cold	2.2%	
2	3	202	187	Nervous-funct.compl.	8.0	
3	4	196	179	Myalgias, fibrositis	9.5	
4	2	183	196	Small injuries		-6.6
5	5	141	141	Comm.cold acc.by temp.	0.0	
6	6	115	97	Adipositas	18.6	
7	7	87	91	Tonsillitis		-4.4
8	8	81	68	Chron.a-spec.resp.cmpl.	19.1	
9	10	68	55	Hypertension	23.6	
10	9	62	62	Acute gastroenteritis	0.0	
11	11	59	55	Cerumen	7.3	
12	12	54	50	Acute bronchitis	8.0	
13	14	52	43	Arthrosis deformans	20.9	
14	13	51	49	Dermatitis (non-occup.)	4.1	
15	16	43	37	Deafness	16.2	
16	15	41	43	Acute otitis media		-4.7
17	17	35	35	Constitutional eczema	0.0	
18	18	34	34	Cellulitis, abscesses	0.0	
19	22	31	26	Myocardial infarction	19.2	
20	21	28	27	Sinusitis	3.7	
In 2000 no longer in the 'top 20':						
21	19	28	30	Oxyuriasis		-6.7
22	20	27	29	Distortions		-6.9

The 20 most often presenting complaints among men in a 'standard GP practice' (n=2800) in the years 1980 and 2000. Concerns only **new** complaints among **men**.

Rank 2000	Rank 1980	abs. nrs. 2000	abs. nrs. 1980	Presenting complaints	percentage incr.	percentage decr.
1	1	281	275	Non-febrile comm.cold	2.2	
2	2	196	179	Myalgias, fibrositis	9.5	
3	2	183	196	Small injuries		-6.6
4	5	146	141	Nervous-function.compl.	3.5	
5	4	141	141	Comm.cold acc. by temp.	0.0	
6	6	86	91	Tonsillitis		-5.5
7	7	62	62	Acute gastroenteritis	0.0	
8	8	59	55	Cerumen	7.3	
9	9	54	50	Acute bronchitis	8.0	
10	10	48	47	Dermatitis (non-occup.)	2.2	
11	11	40	43	Acute otitis media		-7.0
12	12	34	34	Cellulitis, abscesses	0.0	
13	15	28	27	Acute sinusitis	3.7	
14	13	28	30	Oxyuriasis		-6.7
15	14	27	29	Distorsions		-6.9
16	17	24	24	Conjunctivitis	0.0	
17	18	24	23	Dermatophytosis	4.3	
18	16	23	25	Verrucae		-8.0
19	20	22	19	Tendovaginitis, etc.	15.8	
20	23	21	18	Lumbago	16.7	
In 2000 no longer in the 'top 20':						
21	19	20	20	Constitutional eczema	0.0	

The 20 most often presenting complaints among women in a 'standard GP practice' (n=2800) in the years 1980 and 2000. Concerns **all** complaints among women (excl. partus, gravidity, counselling, etc.)

Rank 2000	Rank 1980	abs. nrs. 2000	abs. nrs. 1980	Presenting complaints	percentage incr.	percentage decr.
1	2	333	318	Nervous-funct. compl.	4.7%	
2.	1	323	325	Non-febrile comm.cold		-0.6
3	3	222	197	Adipositas	12.7	
4	4	207	193	Myalgias, fibrositis	7.3	
5	5	158	161	Comm.cold acc.by temp.		-1.9
6	7	141	120	Hypertension	17.5	
7	6	127	131	Small injuries		-3.1
8	10	121	104	Varices	16.3	
9	9	111	106	Ac. inf. of urin.tract	4.7	
10	8	102	108	Tonsillitis		-5.6
11	11	99	83	Arthrosis deformans	19.3	
12	13	74	73	Vaginitis	1.4	
13	12	74	74	Dermatitis (non-occup.)	0.0	
14	14	61	63	Acute gastroenteritis		-3.2
15	15	59	56	Cerumen	5.4	
16	16	56	53	Acute bronchitis	5.7	
17	19	45	38	Deafness	18.4	
18	17	43	43	Constitutional eczema	0.0	
19	18	38	40	Acute otitis media		-5.0
20	20	36	37	Conjunctivitis		-2.7

The 20 most often presenting complaints among women in a 'standard GP practice' (n=2800) in the years 1980 and 2000. Concerns only **new** complaints among **women** (excl. partus, gravidity, counselling, etc.)

Rank 2000	Rank 1980	abs. nrs. 2000	abs. nrs. 1980	Presenting complaints	percentage incr.	percentage decr.
1	1	322	324	Non-febrile comm.cold		-0.6
2	2	213	212	Nervous-funct. compl.	0.5	
3	3	206	192	Myalgias, fibrositis	7.3	
4	4	158	160	Comm.cold acc. by temp.		-1.3
5	5	127	130	Small unjuries		-2.3
6	7	111	106	Acute inf.of urin.tract	4.7	
7	6	101	108	Tonsillitis		-6.5
8	8	73	72	Vaginitis	1.4	
9	9	68	68	Dermatitis (non-occup.)	0.0	
10	10	61	63	Acute gastroenteritis		-3.2
11	11	59	56	Cerumen	5.4	
12	12	56	53	Acute bronchitis	5.7	
13	13	38	40	Acute otitis media		-5.0
14	14	36	37	Conjunctivitis		-2.7
15	16	33	33	Acute sinusitis	0.0	
16	17	32	33	Cellulitis, abscesses		-3.0
17	15	32	35	Oxyuriasis		-8.6
18	18	27	29	Verrucae		-6.9
19	19	25	25	Constitutional eczema	0.0	
20	21	25	23	Adipositas	8.7	
In 2000 no longer in the 'top 20':						
21	20	22	24	Distortions		-8.3

Source: Van den Hoogen et al., 1982, p. 873.

APPENDIX H: DISCHARGE DIAGNOSES OF (OLDER) HOSPITAL PATIENTS, 1980

In this appendix, an attempt is made to give a general overview of the diagnoses encountered among older hospital patients.
The data derive from the most recently published statistics on diagnoses in hospitals 1980 (The Hague, CBS, 1984). The data relates to 95.4% of patients discharged from all hospitals in the Netherlands during 1980.
In cases where the same patient was discharged more than once from a hospital, duplication may have occurred.
The following data are successively presented:

Table H-1: Absolute number of patients according to diagnosis (154 number list), age category and sex.

Table H-2: Number of patients per 100,000 of the population according to diagnosis (18 main groups), age and sex.

Table H-3: Share of the population of 55 and older in the discharge diagnoses (18 main groups), according to sex, compared with age categories 0-14 and 15-54 years of age.

Table H-4: Share of a certain discharge diagnosis (18 main groups) in the total of discharge diagnoses among patients of 55 and older, according to sex (data processed by us).

Table H-1: Absolute number of patients according to diagnosis (154 number list), age category and sex.

Age categories	55-59		60-64		65-69		70-74		75-79		80-84		85		Total	
	M	F	M	F	M	F	M	F	M	F	M	F	M	F	M	F
Infectious and parasitic diseases	303	294	291	280	280	333	287	346	218	355	131	226	88	148	9034	7569
1 Infectious diseases of the stomach and intestines	39	23	43	26	48	32	44	52	47	46	24	46	20	29	1991	1652
2 All forms of tuberculosis	38	22	31	20	30	24	34	39	29	32	19	23	5	4	544	390
3 Certain types of infectious dis.	57	76	55	69	49	86	47	71	26	83	25	60	16	45	2292	1832
4 Other infect. and paras. dis.	169	173	162	165	163	191	162	184	116	194	63	97	47	70	4207	3695
Neoplasms	5255	5848	6336	5341	7905	5710	7493	5755	5705	4474	2937	2673	1468	1417	51902	63644
5 Malig. neoplasms of stomach	239	81	272	120	407	197	346	211	317	231	238	146	118	111	2291	1260
6 Malig. neoplasms of intestines	136	175	229	233	284	334	305	339	268	367	157	262	103	128	1763	2138
7 Malig. neoplasms of rectum	141	168	204	171	273	213	298	277	238	231	148	170	86	99	1641	1491
8 Other malig. neopl. of digest. tract	345	163	412	240	453	310	410	356	317	299	183	239	123	128	2794	2061
9 Malig. neopl. of resp. tract	1583	185	2057	204	2537	233	2229	178	1411	111	498	65	150	27	12298	1370
10 Malig. neoplasms of mamma	2	1156	4	1029	4	1076	12	1025	8	698	5	411	2	195	56	8857
11 Malig. neopl. of cervix uteri		262		283		189		177		92		51		12		3392
12 Malig. neopl. of other parts of uterus		291		272		259		230		154		77		42		1886
13 Malig. neopl. of ovaries and testes	26	438	10	328	14	275	5	210	7	140	4	77	1	23	702	2509
14 Malig. neopl. of prostate	125		310		578		818		829		505		304		3543	
15 Malig. neopl. of bladder	427	48	532	105	693	112	733	188	573	148	333	111	161	66	3886	894
16 Other malig. neopl. of urogen. system	100	125	134	117	167	122	169	169	110	156	43	105	26	49	1043	1280
17 Other primary and second. malig. neoplasms	835	1029	906	953	1052	1007	883	1006	608	820	314	420	163	254	7654	8954
18 Hodgkin's disease	77	60	92	48	69	41	66	80	81	46	13	22	11	18	1079	709
19 Leukemias	146	113	154	95	186	135	168	181	155	136	114	104	46	43	1933	1637
20 Other mal. neopl. of lymph. syst. and flood-form. organs	270	218	226	225	294	269	311	311	169	255	130	139	50	54	2319	2009
21 Benign neopl. of mamma	3	155	-	109	2	108	2	48	-	52	-	23	-	9	35	3572
22 Leiomyomas and other benegn neopl. of the uterus		282		110		89		68		33		16		3		8837
23 Benign neopl. of the ovaries		61		44		53		49		31		11		1		1051
24 Benign neopl. of kidneys and urinary tract	119	25	141	23	164	47	154	49	164	39	60	19	25	11	1008	265
25 Other benign and unspec. neopl.	681	813	653	632	728	641	584	603	450	435	192	205	99	144	7957	9472

Age categories	55-59 M	55-59 F	60-64 M	60-64 F	65-69 M	65-69 F	70-74 M	70-74 F	75-79 M	75-79 F	80-84 M	80-84 F	85 M	85 F	Total M	Total F
Endocriene, nutritional and metabolic diseases and immunity disorders	860	1808	897	1899	946	2389	962	2532	809	2399	526	1536	321	847	11434	22916
26 Diabetes mellitus	527	864	595	1132	666	1530	713	1799	623	1756	396	1154	224	587	6944	12168
27 Atoxic goitre	48	255	43	185	31	162	14	115	9	70	1	24	1	12	341	2187
28 Thyrotoxicosis	56	201	33	178	44	229	40	205	10	158	8	75	3	30	432	2235
29 Other endocrine, nutritional and metabolic diseases and immunity disorders	229	488	226	404	205	468	195	413	167	415	121	283	93	218	3717	6326
Diseases of the blood and blood-forming organs	168	201	264	287	382	411	387	563	415	657	355	513	226	392	4579	5437
30 Vitamin B 12 deficiency anaemia	9	15	16	17	36	38	28	57	34	76	28	50	14	38	182	334
31 Iron deficiency and oth. anaemia	102	128	161	195	227	287	276	401	297	495	264	401	185	322	2404	3469
32 Other diseases of blood and blood-forming organs	57	58	87	75	119	86	83	105	84	86	63	62	27	32	1993	1634
Mental disorders	756	1004	543	734	440	751	340	691	267	527	187	395	147	227	11250	14547
33 Affective psychoses	102	211	80	158	67	155	42	123	20	63	14	23	4	5	786	1675
34 Other psychoses	172	166	142	137	156	224	170	262	177	283	137	295	128	196	2756	3318
35 Neuroses	155	339	116	227	74	186	46	135	22	72	7	22	6	6	2077	4319
36 Mental deficiency	7	4	1	-	-	2	-	3	3	-	1	-	-	-	196	177
37 Other non-psychotic disturb.	320	284	204	212	143	184	82	168	45	109	28	55	9	20	5135	5058
Diseases of the nervous system and sense organs	1950	2122	1823	2005	1938	2469	1926	2721	1537	2464	838	1570	401	708	40163	41123
38 Paralysis agitans	51	58	85	106	158	176	178	244	181	183	62	100	27	18	809	951
39 Epilepsy	156	94	119	81	124	103	119	87	79	74	36	36	17	16	2919	2261
40 Other diseases of the central nervous system	313	285	297	256	271	275	267	243	188	181	90	103	33	48	4237	3957
41 Diseases of the peripheral nerves and ganglia	297	505	227	305	186	319	155	269	99	186	39	95	11	35	5557	5046
42 Strabismus	22	57	18	37	24	29	19	20	10	9	4	8	-	2	3863	4290
43 Cataract	324	299	424	421	584	810	659	1084	602	1202	412	860	206	408	3887	5536
44 Glaucoma	90	105	101	145	108	185	137	209	102	205	46	128	28	65	835	1214
45 Ablatio and other ophthalmic dis.	310	357	308	355	330	410	297	450	242	355	136	225	68	110	4042	4295
46 Otitis media and mastoiditis	148	133	95	127	61	53	35	30	12	26	3	4	3	2	11024	8595
47 Otosclerosis and other deafness	87	116	55	76	28	43	18	24	4	12	5	1	-	1	2013	2120
48 Other dis. of the car and mastoid	152	113	94	96	64	66	42	61	18	31	5	10	8	3	3177	2858
Cardiovascular diseases	12364	5188	12032	5964	12295	7757	11027	8966	8018	8827	4483	6008	2359	3491	91223	63076
49 Acute myocardial infarction	2806	569	2817	819	2822	1200	2502	1476	1676	1408	855	868	377	445	18686	7514
50 Other ischemic cardiac diseases	3524	972	2650	1141	1980	1256	1363	1221	806	957	307	439	121	174	17759	7607
51 Valvular disfunctioning and rheumatic heart diseases	91	130	75	101	42	124	28	57	4	17	4	10	-	7	500	842
52 Myocardial insufficiency and dysrythmia	1287	653	1617	941	2075	1547	2329	2110	2116	2527	1436	1990	862	1357	13942	12407
53 Other cardiac diseases	434	238	378	210	348	297	315	281	193	263	137	173	63	114	3205	2243
54 Pulmonary embolism and infarction	146	103	180	141	239	173	228	231	194	262	101	163	46	79	1555	1608
55 Cerebrovascular lesions	1124	612	1450	890	1825	1332	1914	1791	1569	1979	953	1450	550	794	11342	10319
56 Hypertension	298	355	231	348	212	398	176	340	76	241	38	120	16	37	2291	2846

Age categories	55-59 M	55-59 F	60-64 M	60-64 F	65-69 M	65-69 F	70-74 M	70-74 F	75-79 M	75-79 F	80-84 M	80-84 F	85 M	85 F	Total M	Total F
57 Atherosclerosis	424	87	468	126	545	161	422	217	283	189	156	153	71	104	2979	1241
58 Arterial thrombosis and embolus	349	85	403	127	462	189	325	177	176	152	83	150	62	121	2470	1306
59 Other diseases of arteries and capillaries	960	271	1103	296	1195	361	983	416	575	328	213	176	100	88	6723	2766
60 Phlebitis, thrombosis and embolus	196	152	213	202	224	259	216	294	192	269	117	166	49	87	1739	2152
61 Varices of the lower extremities	342	696	204	401	77	277	49	159	39	83	12	60	7	23	3109	7015
62 Haemorrhoids	258	162	141	138	115	116	73	93	37	38	27	26	10	19	2769	2100
63 Other conditions of the circulatory system	125	123	102	83	134	67	104	103	82	114	44	64	25	42	2154	1110
Diseases of the respiratory system	2537	1374	2634	1394	2946	1483	2987	1390	2331	1173	1329	771	630	453	75953	57606
64 Acute infections of the respiratory system	65	42	78	41	84	44	79	47	83	45	47	33	23	26	3549	2728
65 Hypoertrophy of tonsils and adenoids	34	64	17	53	10	21	4	11	4	4	1	-	-	-	28472	27625
66 Chronic sinusitis	100	78	90	71	55	58	36	41	11	24	4	8	2	4	3798	3066
67 Septum deviation	197	113	118	61	50	33	35	13	10	3	3	1	-	-	9100	4683
68 Other conditions of the upper respiratory passages	380	199	266	171	226	151	131	106	75	48	37	17	11	13	6978	5406
69 Pneumonia, bronchopneumonia and influenza	267	112	304	168	431	190	486	281	455	291	403	260	271	208	5774	3618
70 Bronchitis	439	182	496	234	659	317	752	283	578	275	310	143	108	57	4993	2740
71 Emphysema	214	35	301	41	399	69	395	66	343	53	151	39	47	12	2065	399
72 Asthma and other lung obstruct.	424	398	462	373	507	376	470	334	353	217	169	93	67	37	5819	4878
73 Other conditions of the lower respiratory passages	417	151	602	181	525	224	599	208	419	213	204	177	101	96	5605	2463
Diseases of the digestive system	5755	4506	5289	4062	5048	4431	4476	4412	3398	3980	1938	2947	1166	1812	73599	62333
74 Diseases of the mouth, teeth, salivary ducts and jaws	219	230	130	159	100	120	40	87	42	47	11	23	9	13	3462	3981
75 Gastric and duodenal ulcers	567	255	452	220	394	267	367	278	329	312	182	243	100	183	5603	2788
76 Other diseases of the oesophagus, stomach and duodenum	262	180	242	184	292	189	281	242	214	246	131	233	57	103	4290	3047
77 Appendicitis	218	265	165	184	156	176	93	145	76	76	25	60	18	34	9303	10652
78 Inguinal hernia without ileus	1843	104	1871	103	1676	125	1401	133	923	133	458	69	217	39	19181	3018
79 Other hernias without ileus	379	377	310	366	277	397	218	345	162	251	89	189	38	99	3724	4293
80 Ileus with or without hernia	200	182	211	171	248	253	306	248	246	323	216	327	185	278	3034	2625
81 Chronic enteritis and ulcerative colitis	67	66	49	53	49	47	31	43	20	44	9	8	6	12	1044	1277
82 Fissure, fistula, abscess and other conditions of the anus	216	70	142	50	93	35	66	33	29	17	13	6	5	6	4006	1792
83 Diverticulitis (colon)	175	177	155	214	150	276	152	345	126	362	106	280	47	144	1223	2089
84 Other conditions of colon and peritoneum	408	479	507	481	509	537	534	680	444	681	268	564	217	400	8049	8347
85 Cirrhosis of the liver	246	192	179	170	156	133	95	130	49	69	23	44	14	20	1987	1431
86 Cholelithiasis and cholecystitis	676	1596	609	1439	675	1536	634	1358	533	1111	284	704	182	365	5289	13708
87 Other conditions of liver, gall bladder and pancreas	279	333	267	268	273	340	258	345	205	308	123	197	71	116	3404	3285
Diseases of the genitourinary system																

Age categories	55-59		60-64		65-69		70-74		75-79		80-84		85		Total	
	M	F	M	F	M	F	M	F	M	F	M	F	M	F	M	F
Diseases of the genitourinary system	3252	5462	3397	3714	3735	3349	3558	2718	2875	1831	1745	975	977	454	42069	92481
88 Nephritis and nephrosis	174	132	165	163	169	146	153	125	121	104	89	68	55	42	2102	1583
89 Infections of kidney and/or bladder	74	156	60	155	73	167	75	160	60	127	23	105	24	50	710	2219
90 Calculi of the urinary tract	708	345	505	275	427	237	283	182	173	109	80	52	35	19	5792	3108
91 Other diseases of the kidneys and urinary tract	551	430	574	442	617	486	650	505	572	460	460	309	284	171	7307	7631
92 Prostate hypertrophy	847		1355		1767		1923		1644		945		479		9424	
93 Hydrocele	225		216		202		122		82		30		20		2549	
94 Phimosis and paraphimosis	121		103		92		98		61		41		33		6812	
95 Other conditions of the male genitalia	500		357		316		215		130		65		40		5971	
96 Mastopathia, gynecomastia and other conditions of the breast	52	406	62	221	72	197	39	124	32	65	12	39	7	19	1402	9659
97 Menstruational disturbance and postclimacteric hemorrhage		2002		878		629		490		317		143		63		31163
98 Prolapse of uterus and/or vagina		961		975		1016		832		457		168		44		7920
99 Female sterility		-		-		-		-		-		-		-		5171
100 Other conditions of the female genitalia		1030		605		471		300		192		91		46		24027
Partus and complications of pregnancy, childbirth and puerperium		1		-		-		-		-		-		-		128757
101 Non-spontaneous abortion		-		-		-		-		-		-		-		6507
102 Spontaneous abortion		-		-		-		-		-		-		-		7684
103 Ectopic and molar pregnancy		-		-		-		-		-		-		-		4261
104 Abortus imminens and hemorrhage during gravidity		-		-		-		-		-		-		-		5943
105 Toxicosis, eclampsia and hyperemesis gradidarum		-		-		-		-		-		-		-		7875
106 Other complications of pregnancy		-		-		-		-		-		-		-		27627
107 Partus without complications		-		-		-		-		-		-		-		19003
108 Other indications for obstetrical intervention		1		-		-		-		-		-		-		29565
109 Partus with complications		-		-		-		-		-		-		-		20315
110 Complications of the puerperium		-		-		-		-		-		-		-		1077
Diseases of skin and subcutis	489	531	415	513	386	533	409	748	300	639	252	530	146	342	11463	11423
111 Pilonidal sinus	28	5	12	-	5	2	-	1	-	1	-	-	-	-	1865	721
112 Other infections of skin and subcutis	142	122	124	102	90	103	79	150	65	108	48	71	28	54	3322	2502
113 Eczema and dermatitis	54	52	40	48	52	46	50	47	25	49	29	34	15	24	820	818
114 Psoriasis	53	51	45	47	41	49	45	40	30	41	21	26	8	17	676	630
115 Chronic ulceration of the skin	66	100	86	129	113	174	120	348	117	323	103	307	69	197	986	1856
116 Other diseases of the skin	146	201	108	187	85	159	115	162	63	117	51	92	26	50	3794	4896
Diseases of the musculoskeletal system and connective tissue																

Age categories	55-59		60-64		65-69		70-74		75-79		80-84		85		Total	
	M	F	M	F	M	F	M	F	M	F	M	F	M	F	M	F
Diseases of the musculoskeletal system and connective tissue	4141	4400	2705	3705	1990	3941	1494	3627	961	2729	381	1360	181	510	62543	56125
117 Rheumatoid arthritis	144	325	190	272	138	356	130	350	90	279	26	107	9	38	1139	2493
118 Osteoarthritis and related conditions	335	719	343	909	400	1297	393	1528	277	1172	115	629	46	181	2925	7420
119 Intra-articular derangement	426	328	190	223	106	191	46	91	12	55	6	28	9	16	14476	6512
120 Other arthropathies	191	216	127	160	114	204	90	167	56	167	24	100	15	38	4672	3728
121 Dorsopathies	1736	1193	876	772	504	637	267	449	182	268	59	126	24	46	19558	15475
122 Rheumatism, with the exception of the back	939	738	692	589	507	502	369	405	201	254	76	132	27	47	11884	10046
123 Osteochondropathy	21	28	12	16	2	20	10	11	5	14	2	5	4	-	1368	938
124 Hallux valgus and varus	89	509	55	388	34	379	26	251	16	158	5	81	4	23	861	3952
125 Other conditions of musculo-skeletal system and conn. tissue	260	344	220	376	185	355	163	375	123	362	68	262	43	121	5660	5561
Congenital anomalies	116	119	67	88	69	90	39	59	15	42	14	15	6	4	14326	9840
126 Congenital defects of the nervous system	1	1	1	1	1	1	4	1	1	-	-	-	-	-	353	359
127 Congenital defects of heart and blood vessels	26	23	13	26	14	26	8	12	5	10	-	3	-	1	1648	1493
128 Cleft palate and harelip	5	5	1	2	1	-	-	-	-	-	-	-	-	-	669	409
129 Other congenital defects of the digestive tract	5	5	9	8	9	8	3	8	-	8	3	1	1	1	1305	579
130 Undescended testicle	3		6		1		1		1		2		-		4798	
131 Other congenital defects of the urogenital tract	23	14	16	10	13	19	9	8	5	8	3	1	4	-	1605	1083
132 Other congenital defects	53	71	21	41	30	35	14	30	3	16	6	10	1	2	3945	5917
Conditions arising in the perinatal period	-	-	-	-	-	-	-	-	-	-	-	-	-	-	5807	4469
133 Prematurity and dysmaturity	-	-	-	-	-	-	-	-	-	-	-	-	-	-	1417	1312
134 Other perinatal conditions	-	-	-	-	-	-	-	-	-	-	-	-	-	-	4390	3157
Symptoms and ill-defined conditions	2639	2086	2348	1938	2278	2128	2228	2204	1846	1933	1130	1386	733	911	33067	35750
135 Symptoms	1995	1662	1790	1542	1738	1685	1787	1742	1488	1540	944	1087	645	766	26528	29295
136 Ill-defined conditions	644	424	558	396	540	443	441	462	358	393	186	299	88	145	6539	6455
Injury due to addident and poisoning	2012	2153	1851	2236	1738	2599	1612	3020	1403	3147	1107	2948	908	2804	62216	45873
137 Fractures of skull, spinal column and trunk	278	262	265	257	227	297	209	349	203	348	136	268	100	207	6280	3912
138 Fractures of the shoulder gurdle and upper extremities	125	231	94	277	90	268	60	321	56	308	56	192	35	139	4953	4029
139 Basal neck fracture	128	198	171	304	213	508	262	790	352	1105	400	1453	419	1723	2587	6404
140 Other fractures of the lower extremities	261	347	231	368	169	416	157	371	124	305	93	260	62	210	8869	5304
141 Dislocations and distortions	103	65	72	69	54	70	43	82	34	73	10	46	16	31	5433	2643
142 Cerebral concussion or contusion and other intracranial injury	227	221	223	201	217	243	223	251	166	184	110	117	46	76	10154	5911
143 Internal injury of chest, abdominal and pelvic organs	66	39	60	32	49	22	42	25	28	15	23	16	11	6	2450	860

Age categories	55-59		60-64		65-69		70-74		75-79		80-84		85		Total	
	M	F	M	F	M	F	M	F	M	F	M	F	M	F	M	F
144 Wounds, surface injuries and contusions	281	188	205	171	165	170	136	198	102	203	68	182	66	141	8922	4047
145 Burns	38	35	22	28	15	18	7	19	10	23	8	10	13	11	1599	846
146 Other injuries	57	55	49	49	44	69	33	47	23	32	22	17	8	25	1534	1025
147 Poisoning by drugs or biological substances	111	205	104	165	124	162	104	151	80	190	66	162	55	109	3417	5019
148 Toxic effects of non-medical substances	34	19	21	20	20	14	12	13	11	15	4	7	6	8	1700	1186
149 Complications after medical treatment	297	277	321	290	344	338	321	397	213	340	108	215	69	116	4065	4546
150 Consequences of other external causes	6	11	13	5	7	4	3	6	1	6	3	3	2	2	263	141
Exceptional admissions	1908	1834	1516	1574	1384	1416	944	1245	694	853	329	502	188	323	71351	104702
151 In connection with insemination and development	1	10	2	1	1	1	-	1	-	1	-	2	-	-	1737	39806
152 Healthy new-borns	-	-	-	-	-	-	-	-	-	-	-	-	-	-	44788	42355
153 In connection with specific treatment and follow-up care	894	1205	728	1044	745	918	472	792	371	503	150	304	105	193	14343	13326
154 Other reasons for contact with health care	1013	619	786	529	638	497	472	452	323	349	179	196	83	130	10483	9215
Total	44505	38931	42408	35734	43770	39790	40169	40997	30792	36030	17682	24355	9945	14843	671979	827671

Table H-2: Number of patients per 100,000 of the population according to diagnosis (18 main categories), age and sex.

Nos.	Age group in years									
	<1		1-4	5-9	10-14	15-19	20-24	25-29	30-34	35-39
	born	admitted								
Men	per 100,000 of the male population in each age group									
I	9	894	420	168	72	79	119	126	106	88
II	1	132	124	128	137	130	153	185	201	285
III	7	357	124	63	82	73	84	82	84	109
IV	2	191	132	81	41	36	30	21	14	25
V	-	160	119	61	45	72	172	188	187	225
VI	8	1 052	1 521	1 363	412	297	290	312	328	397
VII	10	106	19	19	48	84	145	268	369	615
VIII	7	2 562	5 827	2 123	478	623	809	666	607	559
IX	4	4 037	1 190	553	502	463	569	664	762	936
X	1	588	734	477	385	290	305	312	326	341
XI										
XII	12	341	162	95	88	180	269	212	176	143
XIII	-	93	158	205	270	661	1 049	1 139	1 284	1 443
XIV	98	2 776	516	648	556	158	114	74	54	54
XV	2 688	3 762	26	3	1	0	0	-	-	-
XVI	16	1 534	582	336	260	196	240	219	251	365
XVII	-	399	1 212	783	742	1 625	1 346	946	790	768
XVIII	49 456	2 631	287	206	168	440	427	328	332	362
Total	52 321	21 615	13 153	7 313	4 288	5 408	6 122	5 741	5 871	6 714
Women	per 100,000 of the female population in each age group									
I	7	744	347	124	67	81	98	87	68	54
II	2	204	126	125	123	183	250	356	564	1 046
III	13	273	102	79	91	104	118	171	187	233
IV	2	138	102	60	38	38	26	29	29	38
V	-	129	87	31	50	135	214	237	250	288
VI	4	824	1 245	1 191	391	316	316	308	355	390
VII	2	63	15	14	23	45	114	220	375	568
VIII	4	1 600	4 587	1 968	629	867	700	523	439	405
IX	1	2 002	729	425	456	604	619	613	637	714
X	-	210	264	282	134	383	992	1 796	2 166	2 620
XI	-	-	-	-	10	867	5 901	9 900	5 106	1 756
XII	4	287	142	80	82	172	226	156	140	141
XIII	2	91	107	139	321	710	700	630	752	1 008
XIV	92	2 155	378	206	197	178	142	145	117	91
XV	2 484	2 829	2	0	0	0	-	-	-	0
XVI	16	1 158	465	295	268	369	436	453	465	535
XVII	-	387	902	492	503	719	516	427	388	449
XVIII	49 236	2 495	224	135	140	257	429	1 502	2 744	2 920
Total	51 869	15 588	9 825	5 645	3 523	6 029	11 796	17 552	14 785	13 256

See Table H-4 for an explanation of the numbers I - XVIII

Tabel H-2: continued.

	40-44	45-49	50-54	55-59	60-64	65-69	70-74	75-79	80-84	≥85	All ages
Men											
I	94	92	80	93	110	126	162	182	197	210	135
II	377	667	1 060	1 609	2 394	3 428	4 223	4 756	4 406	3 505	778
III	163	192	239	263	339	410	542	674	789	766	171
IV	24	36	47	51	100	166	218	346	533	540	69
V	246	273	240	231	205	191	192	223	280	351	169
VI	394	469	549	597	689	841	1 086	1 281	1 257	957	602
VII	1 006	1 811	2 797	3 785	4 547	5 332	6 215	6 684	6 725	5 632	1 368
VIII	549	531	598	777	995	1 278	1 684	1 943	1 994	1 504	1 139
IX	1 089	1 292	1 558	1 762	1 999	2 189	2 523	2 832	2 907	2 784	1 103
X	433	501	706	996	1 284	1 620	2 005	2 396	2 618	2 332	631
XI											
XII	142	146	160	150	157	167	231	250	378	349	172
XIII	1 565	1 511	1 442	1 268	1 022	863	842	801	572	432	938
XIV	46	44	36	5	25	30	22	12	21	14	215
XV	-	-	-	-	-	-	-	-	-	-	87
XVI	438	579	699	808	887	988	1 256	1 539	1 695	1 750	496
XVII	720	648	643	616	700	754	908	1 170	1 661	2 168	933
XVIII	381	428	480	584	573	600	532	578	494	449	1 070
Total	7 667	9 220	11 335	13 625	16 027	18 983	22 639	25 667	26 526	23 743	10 074
Women											
I	64	63	80	83	94	117	140	189	193	194	112
II	1 495	1 859	1 791	1 650	1 783	2 014	2 337	2 381	2 285	1 863	940
III	275	330	456	510	634	842	1 028	1 277	1 313	1 113	338
IV	46	65	56	57	96	145	229	350	438	515	80
V	324	331	322	283	245	265	281	280	338	298	215
VI	430	472	585	599	669	871	1 105	1 312	1 342	931	607
VII	702	937	1 138	1 464	1 991	2 736	3 642	4 698	5 136	4 589	931
VIII	357	374	400	388	465	523	564	624	659	596	851
IX	834	990	1 118	1 271	1 356	1 563	1 792	2 118	2 519	2 382	921
X	2 889	2 963	2 500	1 541	1 240	1 181	1 104	975	833	597	1 366
XI	476	56	4	0	-	-	-	-	-	-	1 901
XII	139	130	158	150	171	188	304	340	453	450	169
XIII	1 155	1 190	1 264	1 241	1 237	1 390	1 473	1 453	1 162	670	829
XIV	61	47	39	34	29	32	24	22	13	5	145
XV	-	-	-	-	-	-	-	-	-	-	66
XVI	526	518	567	588	647	750	895	1 029	1 185	1 198	528
XVII	444	425	516	608	746	917	1 227	1 675	2 520	3 686	677
XVIII	1 874	847	510	517	525	499	506	454	429	425	1 546
Total	12 092	11 599	11 504	10 984	11 929	14 032	16 651	19 178	20 818	19 511	12 222

Table H-3: Share of the population of 55 and older in the discharge diagnoses (18 main groups), according to sex, compared with age groups 0-14 and 15-54.

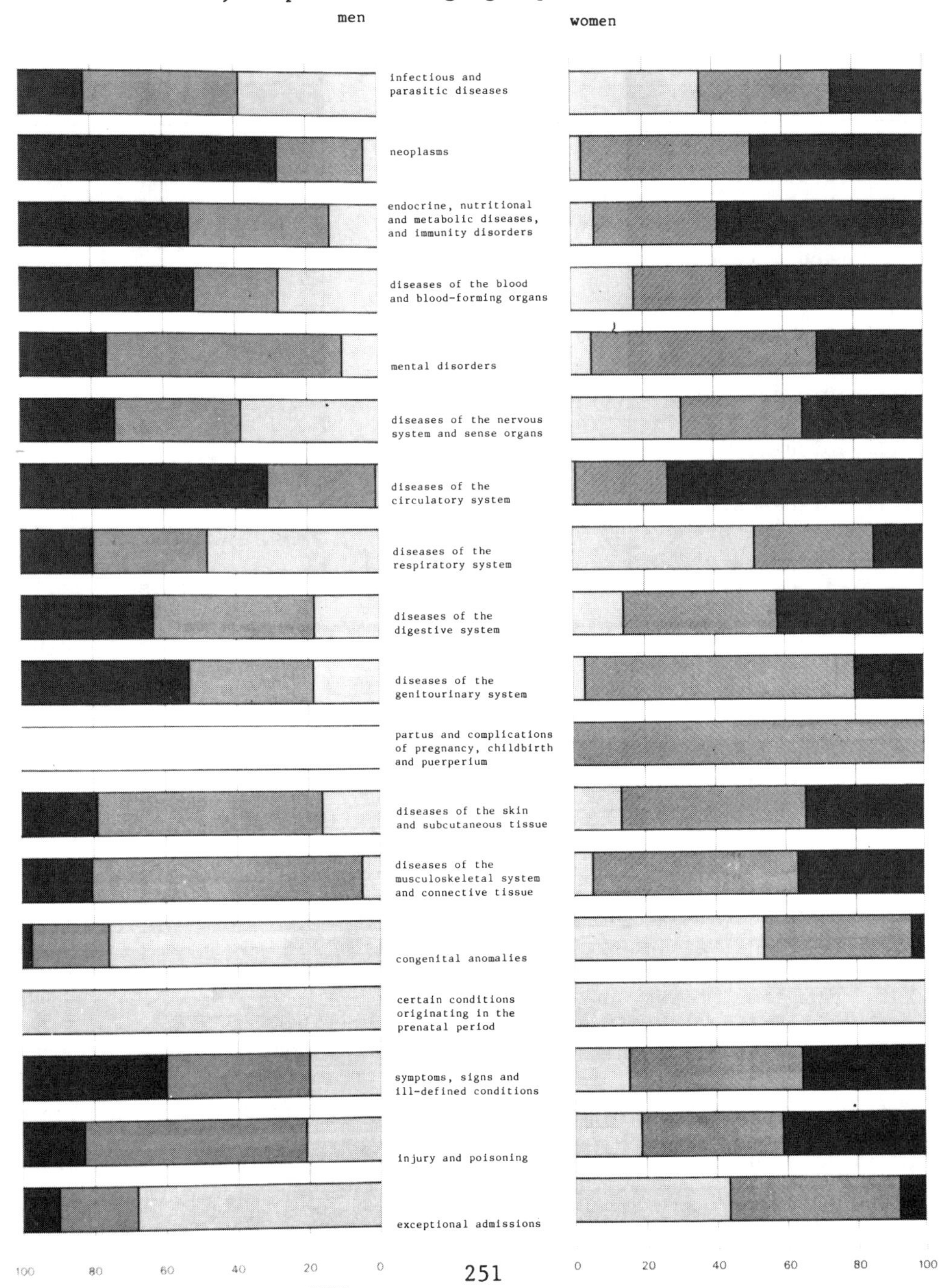

Table H-4: Share of a certain discharge diagnosis (18 main groups) in the total of discharge diagnoses among patients of 55 and older, according to sex.

Discharge diagnosis		absolute		in percentages	
		men	women	men	women
I	Infectious and parasitic diseases	1608	1982	0.7	0.9
II	Neoplasms	37099	31218	16.2	13.5
III	Endocrine, nutritional and metabolic diseases, and immunity disorders	5321	13410	2.3	5.8
IV	Diseases of the blood and blood-forming organs	2197	3024	1.0	1.3
V	Mental disorders	2680	4329	1.2	1.9
VI	Diseases of the nervous system and sense organs	10413	14059	4.5	6.1
VII	Diseases of the circulatory system	62578	46201	27.3	20.0
VIII	Diseases of the respiratory system	15394	8038	6.7	3.5
IX	Diseases of the digestive system	27070	26150	11.8	11.3
X	Diseases of the genitourinary system	19539	18503	8.5	8.0
XI	Partus and complications of pregnancy, childbirth and puerperium	-	1	-	-
XII	Diseases of the skin and subcutaneous tissue	2397	3836	1.0	1.7
XIII	Diseases of the musculo-skeletal system and connective tissue	11853	20272	5.2	8.8
XIV	Congenital anomalies	326	417	0.1	0.2
XV	Certain conditions originating in the perinatal period	-	-	-	-
XVI	Symptoms, signs and ill-defined conditions	13202	12586	5.8	5.5
XVII	Injury and poisoning	10631	18907	4.6	8.2
XVIII	Exceptional admissions	6963	7747	3.0	3.4
Total number of diagnoses ($\geqslant$ 55 years of age)		229271	230680	100.0	100.0

APPENDIX I: FINAL REPORT OF THE CONSULTATIONS

In May 1984 the scenario commission on ageing and the research team organized six discussions with groups of experts from the areas medical and medical-technological research and medical practice. Participants were selected in close consultation and co-operation with the Royal Dutch Society of Medicine and the Steering Group for Research on Ageing.
The consultations were intended to give a wider foundation to pronouncements in the scenarios on medical and medical-technological developments between 1984 and 2000 than would be possible on the basis of information deriving from the literature and knowledge of members of the scenario committee. The participants were asked to check the reports of the consultations.
On the basis of their reactions, one final report was written.
A list of names of participants follows in the final report.

INDEX

I HEALTH SITUATION OF THE POPULATION IN GENERAL

1. Absolute and average life expectancy

The participants in the group discussions expect no shifts with respect to maximum life expectancy (at present approximately 115 years). Of greater importance is the average life expectancy. This is at present 75 years, and an increase of 5 years is to be expected within a few decades, though this increase is significantly less than the leap forward after the thirties and forties.
The gap in life expectancy between men and women (at present 7 years at birth) will remain constant for the present generation of elderly. In coming generations it is expected that there will be a narrowing of the gap, partly brought about by a greater similarity in lifestyles of men and women.

2. Developments with respect to disease

For the future, diseases with a long incubation period (including some types of cancer, coronary and arterial diseases, dementia and diseases brought about by persistant viruses) will continue to play an important role. In the case of increased life expectancy and improved possibilities for treatment of diseases not known, it is expected that so-called 'new diseases' characterized by an even longer period of latency, will make their appearance.

The great disadvantage of existing diseases with a long period of latency is that for coming generations the **prevalence** of the diseases will alter little; in other words, if no drugs are discovered for curing the diseases, assuming similar lifestyles and dietary patterns, the prevalence of disease will remain constant in the coming generation. This implies that in future, on account of double ageing of the population, there will be an absolute though not relative increase in the prevalence of disease.
For coming generations a shift might come about in the prevalence and incidence of disease if there were an alteration in lifestyles and dietary patterns. There is little insight into lifestyles of either the young or the elderly, as a consequence of which it is not possible to make forecasts for the distant future.

As regards cure or retardation of (the most important) diseases, no breakthroughs of any significance are expected for reasons which will be dealth with in Section III. Expectations are that for the time being only very slow progress will be made per specific disease. For the future, emphasis will be on palliation, and the resulting improvement in the quality of life. Possible breakthroughs are only to be expected from preventative measures (preventative drugs or alterations in lifestyle). The outlook is nonetheless pessimistic with respect to prevention in view of the fact that it is extremely difficult to introduce change in people's behaviour. Also, in view of the emphasis on diseases with long incubation periods, the effects on the health situation may only be expected after some generations. In addition to preventative measures, expectations focus strongly on early diagnosis. This entails regular tests and controls with relatively simple aids for discovering diseases in an early stage so that more effective therapy will be possible.

In addition to the prevalence of specific diseases, multimorbidity with possibly a deterioration of mental capacities among the elderly will play an important role. The elderly can increasingly and also at an increasingly older age profit from medical treatment thanks to medical and medical-technological innovations. Some therapeutic and diagnostic methods will however prove to be too expensive to be applied on a large scale. In addition to the costs aspect, the physician must increasingly decide against treatment in view of the poor psychosomatic condition of the patient (multimorbidity and/or lessening of mental capacities). Thus in spite of improved methods of treatment, this dilemma prevents a further improvement in the health situation of the elderly. In addition to this dilemma, there is the possibility that if, thanks to better treatment, the older citizen can remain longer in his own environment, it will be necessary for social conditions to be met in order to offer the elderly a worthwhile, independent existence. (Intensification of first echelon facilities, resocialization, worthwhile use of time, etc.).

II FUTURE DEVELOPMENTS IN FUNDAMENTAL RESEARCH AS TO THE PROCESS OF AGEING

General

In addition to research as to causes and combating of specific diseases, fundamental research as to the process of ageing plays an important role. In the process of ageing a 'natural degeneration' of cells take place, as a result of which the chance of contracting diseases increases. With increased insight into the process of ageing, it will be possible to trace in how far this degeneration is normal or pathological and whether normal and pathological ageing processes can be influenced (influencing of hereditary traits, immunology etc.). It should be stressed that multidisciplinary research is essential to progress.

By comparison with research as to the combating of disease, only a very small budget is made available for fundamental research (1%). This is viewed as one of the most important reasons why, within the time-limit set for the scenarios, no breakthroughs of any significance are to be expected.

1. The problem of choosing models on which to base research on ageing

The big problem in analyzing the ageing process is the choice of a model on which to base research as to the process of ageing. It is posed that a dynamic model must be chosen. The deterministic model which assumes that a certain defect which becomes manifest at a later age is genetically determined at an early age must be discarded on the grounds that it is at odds with the evolution theory. To maintain the existence of a certain species it is essential that there should be variations in the genome, which is why the genome is constantly being reshuffled to obtain the variation necessary for maintaining the species. A dynamic model thus fits in better with the evolution theory.
In addition to the question whether or not a dynamic model should be chosen, a second problem was put forward, namely the unavailability of an adequate animal model. For the purpose of analyzing the process of ageing, especially of the brain at the level of the cell,

it has not been possible to find adequate animal models for pathological studies (for instance of senile dementia) since in the case of man it is often the higher functions of the brain which are first impaired.

2. (Molecular) biological research

This branch of research concentrates mainly on ageing processes at the level of the cell, and especially the transfer of information between cells and the hereditary characteristics of cells.
For the future, it is expected that better understanding will be achieved regarding deviations of a somatic or hereditary nature at the genetic level. With the aid of DNA research it is hoped that it will be possible to map the genome, with the result that it will be possible to detect defects in an early stage. In addition to this, immunological and neurobiological research are also of great importance.

DNA research

It is expected that in the very near future it will become possible to map the human DNA. Mapping the genome would open possibilities for determining (preventative) risk factors for certain diseases. However, the influence of the social environment and the variability of the genome should not be underestimated. Consequently possible preventative (prenatal) screening is not expected to exert much influence on the health situation of the population.

Immunology

One of the body's most important defence mechanisms is the immune system. It is known that in the course of ageing changes occur in the immune apparatus. In this system, an important role is played by substances which regulate the transfer of information between the cells. It is expected that it will become possible to isolate these substances in their pure form, which will open possibilities for regulating the immune apparatus. To start with, this regulation will be a-specific, and will be confined to pepping up weaknesses in the immune system. It is however expected that at a later stage specific regulation of the immune system will become possible, inter alia for intervention in the event of the formation of autoaggressive antibodies. One application of this would be the combating of autoimmune diseases which are caused by certain cells. The significance of this development for the elderly is however limited,

in view of the small number of elderly people with pathogenic antibodies. Non-pathogenic autoantibodies may be considered normal among the elderly. The significance of these antibodies is as yet unclear.
More is expected from the application of knowledge of the immune system to other areas, such as endocrinology. In this context, the linking of toxicants to antibodies which can then be directed at receptors, or making antibodies which can combat a too high concentration of hormones, is envisaged. Another area with which immunology could co-operate is psychiatry. Neuropeptides are a recent result of this co-operation.
Immune deficiency is not considered to be among the causes of cancer though it does play a role in combating it, mainly in the form of immunotherapy which, as we already said, consists of the exogenous administration of antibodies with toxicants aimed at specific cancer tumors.
It may thus be posed that in the coming years possibilities will be opened for coming to grips with the process of ageing at the level of the cell, especially by means of intervention in the immune system. Applications may be expected in the combating and healing of diseases, and especially diseases with a long incubation period.

Neurobiological research
Within this discipline, research as to changes in neurons during the process of ageing and in dementia plays an important role, for instance with the aid of immunocytochemical markers (for example measuring changes in synaptic membranes, neurotransmitters or enzymes).
The relation between normal and pathological ageing calls for attention in this context, especially as relating to Senile Dementia of the Alzheimer Type (SDAT). It may be expected that in the coming years more insight will be obtained into the systems which show primary changes during the process of ageing and SDAT. Such observations might offer the key to early diagnosis and a rational therapy for various forms of dementia.

3. Pharmacological developments

There was some difference of opinion among participants concerning the question how long it will take before drugs which have been found to be effective in animal models will become available for general use. Estimates ranged from 3 to 5 years for modifications of existing drugs, and from 5 to 10 years for completely new drugs.

Little hope was entertained of pharmacological breakthroughs for the purpose of healing disease. On the one hand, the pharmaceutical industry is dependent on insights deriving from the medical world with respect to breakthroughs in knowledge as to the causes of certain diseases. But the pharmaceutical industry is also dependent on the economic yield of its products (investments in research, toxicological research, clinical trials, etc.).

Consequently it will not be possible to devote attention to every disease, but attention will have to be limited especially to those diseases from which a large part of the population suffer.

III FUTURE DEVELOPMENTS FOR THE SPECIFIC DISEASES

Introduction

In this section an indication will be given per disease of expectations with respect to treatment methods and prevalence. Naturally, it was not possible to discuss all diseases. The following pages report on all the diseases discussed, either summarily or at length.

1. (Malignant) neoplasms

Malignant disease accounts for some 25% of total annual mortality. Among men cancer of the lung and among women cancer of the breast account for the greatest share of mortality.
Many people think that the chance of developing cancer has greatly increased compared with forty years ago. This belief is incorrect. It is true that there has been a considerable increase in the absolute number of cases of cancer, but this is brought about by the sharp increase in the size of the population, and especially the sharp increase in the number of elderly persons (as a consequence of the increase in average life expectancy). For women, compared with forty years ago, the chance of developing cancer has even decreased due to the sharp decrease in cancer of the stomach. Among men the chance of developing cancer has increased slightly as a result of a higher incidence of cancer of the lung. The chance of dying from other forms of cancer has however remained constant. There is still much uncertainty as to the causes of cancer. Among the things which are thought to have contributed to the decreased mortality from cancer of the stomach are the falling into disuse of pickling of foods, the advent of refrigerators, and the large-scale availability of fresh fruit and vegetables. Comparative studies have shown that dietary habits as well as other aspects of lifestyle would appear to play an important role in the prevention of cancer. Not so much the surrounding environment but rather the personal environment (food, sexual habits etc.) would appear to play a crucial role.
As in the case of dementia, there is criticism of the registration of morbidity and mortality arising from cancer. Especially the regional differences found in mortality from certain types of cancer are questioned. Improved registration of incidence is necessary so that better statistical material will be available for testing hypotheses with respect to trends in mortality resulting from cancer.

Prognoses remain poor with respect to future possibilities for curing malignant tumors. No general drug will be discovered since tumors, as opposed to viruses and bacteria which cause infections, are not foreign bodies. The organism itself causes the growth. Consequently no general therapy for the cure of cancer may be expected. The tissue specific to each type of tumor calls for selective diagnosis and each case requires its own plan of treatment.
Since the Second World War, about ten cytostatics have been developed (for instance cisplatinum for cancer of the testis). New medicines may be expected in the future, but these will rather be the result of trial and error than consciously compounded drugs for combating cancer. As far as therapy is concerned, expectations are thus fairly negative. Thanks to better surgical techniques, it will in the future be possible to treat benigh tumors more successfully. Moreover increasing attention to after-care is expected, especially as far as this concerns rehabilitation and resocialization.

Preventative measures could yield considerable results. If for instance, everyone were to stop smoking, mortality resulting from cancer would decrease by 30%; if people were to adopt more healthy dietary habits and lifestyles, there could be further decrease of 35% in deaths from cancer. However, in view of experience up to the present of reactions to knowledge regarding the effects of smoking, alcohol and the consumption of fats, expectations that a more healthy lifestyle will be adopted are limited.

Another preventative measure is mapping of the genome, which would make it possible to place people in risk groups at an early age. The disadvantage of this method is however that it is not yet known of many sorts of cancer whether they are hereditary.
Expectations for influencing heredity are negative. Moreover, the variation of the genome could have a disturbing effect (see fundamental research). The sum total of this is that screening at an early age is not expected to exert any significant influence on the prevalence of cancer.

With the available apparatus, it is now possible to make good diagnoses. Advanced apparatus such as NMR will not improve diagnosis, but could be employed in the monitoring of developments in tumor tissue and the effects of drugs on tumors.
Immunological methods (for instance tumor imaging with radioactive antibodies) could prove valuable in the future. There are ongoing developments in this field, and the first results are encouraging.

Finally, little is expected of mass screening. The regular spontaneous regression and disappearance of tumors, faulty diagnosis with all the unpleasant consequences for the patient, the rapid growth of especially mammary tumors which necessitate very regular control, are all arguments for discarding this approach.*)

2. Endocrine, nutritional and metabolic diseases

Studies have shown that there have been radical shifts in dietary habits. Characteristic of these changes are the increasing use of fast foods, increasing consumption of fats, substitution of dietary calories by alcohol calories, over-consumption and a shortage of vitamins.
These shifts result in a considerable deterioration in health, and will lead to a sharp increase in metabolic diseases.
Expectations for the future are pessimistic. Preventative measures based on geared changes in dietary habits (more roughage, less fats, low cholesterol, less smoking etc.) are possible but difficult to realize.

Diabetes
Better treatment is expected for this disease. Pancreas transplantation and beta cell transplantation will however not result in the degree of progress which was expected of them.

*) The scenario committee would like to add that recent data from various projects for mass screening for cancer of the breast and cervical cancer do give strong indications that this approach does result in a significant reduction in mortality from these cancers.This claim was put forward with respect to cervical cancer during a meeting of screening experts under the auspices of the World Health Organization in Copenhagen in 1983 and in a meeting of the European Committee in Luxembourg in 1983, where data from a number of countries were analyzed. With respect to cancer of the breast see for instance A.L.M. Verbeek et al., 1984; H.J.A. Collette et al., 1984; and J.D.F. Habbema et al., 1983.

Intestinal diseases
While ulcerations of the stomach and stomach cancers are on the decrease, a further increase of polyps of the intestines is expected. Endoscopy and lasers however offer good prospects for treating these non-invasively. These methods are also applied in the case of small intestinal hemorrhages resulting from small arterial lesions. It is expected that in the future these methods will also be applied in the treatment of tumors of the colon.

Liver
As a consequence of the increased consumption of alcohol, a clear increase in the number of liver cirrhosis patients is observable. The range of this increase is comparable with France. Prognoses for the treatment of this disease remain poor.
Another problem is possible long-term damage to the liver (7-15 years) resulting from the use of a combination of medicines. In this framework, and partly in connection with the increase in the injurious side-effects of drugs among the elderly, there are pleas for centralization of data from pharmacists regarding the issuing of drugs to patients so that possible harmful interactions can be avoided. Moreover, in the course of development of new drugs, more research should be devoted to possible interactions with other drugs.

Pancreas
In addition to an increase of cirrhosis of the liver, a sharp increase is also observable in cancer of the pancreas which at the moment results in unsuperable diagnostic and therapeutic problems. This increase is probably brought about by lifestyle, especially the increase in the use of tobacco and alcohol.

Gallstones
It will probably become possible to treat this disease with the use of non-invasive endoscopy. A method similar to that applied in the case of kidney stones (pulverization) is however not expected since the gallbladder if often in such poor condition that such treatment would be too risky.

Thyroid abnormalities
These abnormalities are strongly related to age, and expectations are high that early screening will be effective.

3. Diseases of the nervous system and the sense organs

Cerebrovascular diseases
These will be encountered less in the future.

Parkinson's disease
The prognosis for the cure of Parkinson's disease is less optimistic than fifteen years ago when the drug Dopa was introduced. Some success has been achieved with the transplantation of embryonic brain cells for replacing brain functions which have been destroyed. Moreover it is stressed that relatively many Parkinson patients develop dementia (the Parkinson dementia complex). Future embryonic transplants would have little effect in this form of Parkinsonism (multimorbidity) due to the large number of damaged (brain) functions.

Sight
It is expected that both cataract and glaucoma will disappear provided proper diagnosis is made. The outlook with respect to degeneration of the macula lutea is however pessimistic. The frequent occurrence of deterioration of sight (in its most serious form blindness) among elderly diabetics has often been stressed. See Section IV Medical-technological developments for application of the artificial eye.

Hearing
Partial or total loss of hearing involves serious psychic and social implications (withdrawal from society, loneliness, isolation etc.). Up to the present little attention has been devoted to this affliction. It is of a strongly degenerative nature and it is expected that there is little hope for treatment in the future.

Taste
Generally speaking, at a later age there is a deterioration of the sense of taste till eventually the individual can only distinguish sweet and salt. These correspond with foods which often have a detrimental effect on the older individual (teeth, heart and blood vessels, overweight, etc.). Other taste is lost, so that there will possibly be a lack of appetite, while the chance of food-poisoning will increase as a consequence of eating decomposing food, and vitamin deficiency may arise (cf. the possible influence of vitamins and minerals on dementia). One solution might be the addition of a

heightened dose of flavouring substances to foods so that the elderly can longer enjoy the sense of taste. This too however entails unfavourable aspects.

4. Coronary and artirial diseases

In the case of coronary and arterial diseases, surgery is applied to increasingly older people (now 70 to 75 years of age). It is expected that in the future the age criterion for treatment will disappear altogether. Only for preventative surgery is it expected that the limit for treatment will be set around the age of 70. A sharp increase in these diseases is expected as a consequence of the future demographic structure of the population. It is characteristic that till approximately the age of fifty, mainly men die of coronary and arterial diseases. Above this age male surplus mortality gradually declines. It is expected that this surplus mortality will slowly disappear.

Treatment
Palliative measures occupy an important position in the treatment of coronary and arterial diseases. Operative and some other treatments are often not aimed primarily at prolonging life, but rather at improving the quality of life of the patient. One exception is heart transplantations which are only performed on patients with a life expectancy of less than one year, while after transplantation approximately 50% of the patients survive for five years.
There is no general consensus of opinion regarding the introduction of heart transplantation in the Netherlands. Most participants however consider the method so important that they are convinced that it should be introduced.
It is stressed that only a small group of patients (a maximum of a few dozen per annum) should be eligible for this new treatment. Of course the availability of heart donors will play an important role. Moreover it should be pointed out that as far as the elderly are concerned, only in very exceptional cases is heart transplantation a suitable method of treating heart defects.
Prospects with respect to the artificial heart are much less optimistic. The enormous amount of energy consumed by this apparatus continues to constitute a difficult problem. It is consequently expected that the artificial heart will serve only as a provisional solution till a donor heart becomes available. Methods of dealing with the clotting problem will continue to improve.

Diagnosis
The importance of early diagnosis has been greatly emphasized. The tendency to invest in and employ increasingly more expensive apparatus was strongly criticised. More benefit is expected from regular screening of risk groups with simple apparatus.

Future developments
- High hopes are entertained with respect to anticoagulants. With the aid of the recombinant DNA technique, it should be possible to manipulate bacteria genetically for this purpose.
- Balloon catheters (are already employed).
- Laser beams directed through cardiac catheters for enlarging blood vessels and the introduction of better substances than those presently employed, for the removal of clots. The problem involved in the use of laser is to aim the laser beam through the predominantly twisted blood vessels.
- There will be much progress, both qualitatively and quantitatively, in electroregulation of the heart. More complicated, but also more physiologic forms of pacemakers are gradually being introduced for the treatment of heart block, but the treatment of other rhythmic disturbances by means of electroregulation is still in its infancy.
- Artificial heart and heart transplantation.

5. Diseases of the urinary tract

Incontinence

Both physically and socially, incontinence constitutes a serious problem. Greatly improved diagnosis of incontinence is necessary to distinguish clearly its various forms.
Four causes of incontinence are quoted:
a. defects of the organs (prolapse etc.); surgical intervention offers good prospects
b. neurological defects; these are very difficult to treat
c. obstruction (deviation of the urethra); in the future there are good prospects for treating this form of incontinence
d. instable bladder (e.g. in dementia); very difficult to treat.

As in the case of other diseases, there will be enlarged possibilities for treating incontinence at an increasingly older

age. In view of multimorbidity and the deterioration of mental capacitites, the limits of medical treatment are increasingly determined by the mental and physical condition of the patient. For instance, in the case of incontinence it would be possible to implant advanced aids such as artificial muscles and alarm systems, but due to poor functioning of the central nervous system in elderly people, it is often necessary to refrain from applying such treatment and to resort to (socially often less acceptable) non-invasive aids such as diapers, catheters etc. These methods involve the risk of infection.
In the future there will be increasingly better methods of dealing with incontinence, strongly focused on improvement of palliation by means of closer co-operation with other medical fields such as internal medicine, psychiatry, anesthetics, as well as improved social aid, district nursing etc. A combination of new techniques, improved assistance, after-care, rehabilitation and resocialization are thus expected to result in better ways of dealing with this problem.
After-care is at the moment extremely problematic.
Ideas were put forward for a new specialism: pelvic medicine. This specialism would unite the existing specialisms of urology, surgery, neurology and gynaecology, also with the necessary psychological guidance (risk factors as a consequence of events in one's lifetime) in order to achieve better treatment of the diseases involved among the elderly.

6. Diseases of the skin and subcutaneous tissue

Decubitus

Improvements are expected in the area of prevention and treatment of decubitus, but there will be no fundamental breakthroughs.

7. Diseases of the musculoskeletal system

Rheumatic complaints

In a recent epidemiological survey carried out in the Netherlands, 30% of the interviewees claimed to suffer from rheumatic complaints. Similar figures were also found in other western countries. This means that there are 3 million people in the Netherlands with

rheumatic complaints. Of these, 15% (450,000) are undergoing medical treatment. There is a vast discrepancy between the prevalence of rheumatism and the use made of medical facilities.
Since the most important chronic and disabling rheumatic diseases show an increasing prevalence in the older age groups, the need for rheumatological care will increase sharply in the future as a consequence of the ageing of the population. This is also apparent from the present increased medical consumption for locomotor complaints and the increasing use of antirheumatic drugs with increasing age.
Apart from the demographic factor, there is no autonomous trend observable towards a decrease or increase of rheumatic complaints.
Some rheumatic diseases are of a chronic nature and result in lengthy invalidity. The most important are:

- rheumatoid arthritis, encountered among approximately 175,000 Dutch people;
- arthrosis and spondylosis, which mainly arise at a fairly advanced age, and which will eventually afflict 3 million Dutch people; annually about 60,000 people appeal to the rheumatologist on this account;
- the so-called seronegative spondyloarthropathies from which approximately 20,000 people suffer.

The surgical replacement of joints is an important aspect of treatment of rheumatic complaints at a later age. Adequate treatment may result in a considerable reduction of complaints and invalidity. For the time being, it is not possible to indicate developments in the sphere of treatment and prevention of rheumatic complaints which might exert a positive or negative influence on the prevalence of rheumatic diseases in the period 1984-2000.

Future expectations with respect to rheumatic complaints may be summarized as follows:

- the number of rheumatic complaints will increase due to ageing of the population;
- under-diagnosis of rheumatic complaints will decrease;
- GPs' understanding of the disease will increase as a consequence of more attention being devoted to it in training;
- scientific research will increase.

Taken together these developments will result in an increasing need for rheumatological care.

Osteoporosis

Data on the number of patients admitted to hospitals in the Netherlands in the year 1972, 1977 and 1982 show that among women of 65 and older there is an increase in the number of fractures of the upper-arm and forearm, femoral neck fractures and osteoporosis. Among men of 65 and older too there is an increase in the number of femoral neck fractures and osteoporosis; in the case of other fractures this increase is absent or much less striking. In 1972 slightly over 5700 people of 65 and older were admitted for femoral neck fractures or osteoporosis. In 1982 this number had climbed to nearly 10,200. On the basis of the increase over the years 1972-1982 and the expected development of the population, it has been calculated that in 2010 2.5 times as many patients of 65 and older will be admitted on account of a femoral neck fracture or osteoporosis. On 31st December 1982 fractures and complaints involving the spinal column (exclusive of arthrosis) were the most important factors in invalidity in 10.5% of patients of 65 and older resident in nursing homes. It is expected that there will be an increase in this percentage in the future. To maintain a satisfactory level of activity among the gradually ageing population it is essential that research should be carried out as to bone metabolism and the function of the joints and muscles.

In additon to regular screening, substitution with oestrogens and progestogens as preventative measures are recommended. However, this treatment also entails disadvantages, and thus cannot be employed in the case of all patients with osteoporosis. Fluoride stimulates the manufacture of bone, but this medication too has a number of drawbacks. In recent years, calcitonin has been used on a fairly wide scale in many countries, but without sufficient clinical research. Almost nothing is known of the effect of exercises on the skeletal framework. They would however appear to exert a positive influence and diminish chances of falling.

Treatment of osteoporosis is clearly in the process of development and requires express support from scientific research.

The prevalence of osteoporosis is so great that for this region alone it is necessary to seek after methods of treatment, and preferably such as can be administered by the GP.

8. Mental disorders

Dementia

a. Etiology

It is expected that after 2000 there will be vast increase in the understanding of the etiology of dementia resulting in considerable breakthroughs in this area. Till such time, it will be necessary to concentrate on palliative measures; that is to say, on the basis of early diagnosis, to attempt to halt or delay the process of the disease by means of adapted selective therapies.
The general question in research as to the cause of dementia is what changes exactly take place in the brain, and whether these changes are the consequence of an accelerated physiological process of ageing or the result of pathological processes. In this context, one of the most urgent problems is the need for more systematic recording of dementia symptoms so that it will be possible to arrive at a better differentiation between the various types of dementia. Only when it is clear what specific sorts of dementia exist, will it be possible to look for the specific processes which result in this state. Moreover, with much better diagnosis, therapy can be attuned to the disease.
Research as to the etiology of dementia can focus on various fields:

- research with animal models

With the aid of animal research, an attempt is made to draw up a model which makes it possible to understand dementia. The obstacle which this approach presents however is that up to the present no adequate animal model has been found. One of the reasons for this is that dementia affects primarily the higher functions of the brain.

- postmortem examination

By means of autopsy, an attempt is made to obtain insight into the changes in the brain structure brought about by dementia. Postmortem examination is hardly performed in the Netherlands.

- studies as to the relationship between diet and dementia

In this area an attempt is made to obtain some insight into the relationship between vitamins and minerals and the functioning of the brain. Vitamin deficiency might possibly result in impaired cognitive functioning.

- hereditary research

This area of research is focused on the relation between dementia and (possible) heredity. Attention is drawn to Pick's disease which might be hereditary.

- neurobiological research

Research in this area is directed at changes in neurons during the process of ageing and dementia. In this context, the relation between normal and pathological ageing calls for attention, especially as this relates to Senile Dementia of the Alzheimer Type (SDAT). It may be expected that in coming years more insight will be obtained into the systems which evince primary alterations during ageing and SDAT. Such observations might provide the key to early diagnosis and a rational therapy for dementia.

It is furthermore stressed that research as to the causes of dementia should be supported by longitudinal epidemiological studies with university backing, in order to obtain better understanding of the prevalence of the disease in the Netherlands. Data now employed are often extrapolations of figures from countries with a comparable state of health.

In addition to this, co-ordination of research (neuropathology, neuropharmacology and neuropsychology) as well as organization of the availability of research material (both old laboratory animals and human brain tissue) are essential prerequisites for successful research as to this disease.

If ground is to be won in the battle against dementia, it is very important to know whether certain forms of the disease are reversible. Opinions of the participants in the consultations varied considerably on this point. Some experts were of the opinion that some part of the brain function is lost and cannot be restored. Complete cure of these forms of dementia would thus be impossible. Other experts argued that these forms of dementia are reversible; in the future a drug will be found with the potential of restoring lost functions. Total cure should thus be within the limits of the possible.

b. Diagnosis

Attention was repeatedly drawn to the importance of early diagnosis of dementia. If dementia is diagnosed in an early stage it is often possible to retard the process considerably with the aid of adapted therapies. In the view of participants of the consultations, in order to be able to diagnose the disease at an early stage, it will be necessary to acquaint a wider public with its symptoms so that people in the vicinity of the patient will sooner recognize the disease. Among first echelon facilities too, (e.g. GP) a wider knowledge of the disease is much to be recommended, so as to awaken a greater alertness to possible sign of dementia among older patients. A wider knowledge of the disease might be acquired by means of post-graduate courses, the introduction of a course in geriatrics (as a compulsory part of training), etc. Of great importance too is the distinction between dementia and pseudodementia (often caused by drug intoxication). Expectations are that improved diagnosis so that the disease is recognized at an earlier stage could serve to delay it by 4 to 5 years. After the year 2000 it is expected that possibilities for treating dementia would have improved.

c. Therapy

Till such time, with the aid of adapted therapies, every effort must be made to delay or halt the process of the disease. There are various ways in which the disease might be delayed. One method is the administration of trophic pharmaca (neuropeptides and neurotransmitters) to stimulate brain function or to prevent further deterioration. In addition to this, experiments are being carried out with drugs which improve the functions of memory and speech. Up to now this latter method has met with little success: the present drugs rather act as placebos. It is however expected that there will be breakthroughs in this area in 5 to 10 years.
Training programmes and occupational therapies are employed to create an enriched environment which it is hoped will help delay accelerated ageing of the brain. Moreover, the importance of a good diet containing a high level of minerals and vitamins for improving the cognitive functions and reducing cases of depression is stressed. The elderly are apparently very sensitive to an imbalance in vitamins etc. and have high requirements for these substances.
Apart from these existing - successful or less successful - therapies, the possibility exists that in the future transplants will be carried out with embryonic brain cells to compensate the loss of the patient's own cells resulting from disease. Greater success is however expected in the case of diseases such as Parkinsonism than in the case of dementia in view of the wide range of disturbed functions in the latter disease.
Consequently, any important neuropharmacological breakthroughs in the short term would seem unlikely, since many neurotransmitter systems evince changes which in most cases indicate a heightened or accelerated process of ageing.

d. Expectations with respect to dementia

- a sharp increase in the incidence of dementia in view of double ageing;
- more systematic recording of the symptoms of the various forms of dementia;
- after 2000 clear insight into the etiology of dementia;

- as a consequence of this insight, specific drugs for the cure or prevention of dementia;
- better treatment possibilities, both pharmacologically and psycho-socially;
- better attunement of drugs for dementia patients and possibilities for monitoring (with scanning apparatus) the changes in brain tissue;
- improved possibilities for early diagnosis of dementia, which would result in vastly improved results from treatment.

9. Teeth

It is expected that there will be a considerable increase in the number of elderly persons who retain their own teeth.

IV MEDICAL-TECHNOLOGICAL DEVELOPMENTS

General

Medical-technical apparatus serves to aid the physician in making diagnoses and in implementing and monitoring therapy.
Technical apparatus continues to improve, but it also becomes increasingly more complicated and expensive (think of NMR which is costly to purchase, install and use). In the future, the cost factor will play a crucial role in the development of technical apparatus.

From a technical point of view, possibilities already exist for solving all practical problems, but only at high cost. Moreover it may be questioned in how far these relatively small improvements and refinements weigh against the costs of the new apparatus. As against this, once new apparatus has been introduced, large-scale production might result in making it considerably cheaper. As far as development of technical aids is concerned, it is expected that there will be a vast increase in innovations aimed at the elderly as a result of the application of microelectronics. These aids often do not serve to prolong life, but considerably improve the quality of life of the (handicapped) patient, and make it possible for him to remain longer and more independently in his own surroundings.

One problem posed by this new apparatus is that it becomes increasingly complicated and compact, which might involve operational problems especially among the elderly (think of people with a tremor). The application of ergonomic principles could however greatly facilitate the use of this apparatus by the elderly.

Possible future developments

a. Medical-technical apparatus

In the area of medical apparatus it is expected that better scanning techniques will be developed for making diagnosis and monitoring the course of therapy.
Lasers will be widely employed in heart and eye operations (and in general in other small operations).

b. Aids

This refers among other things to the development of prostheses. It is expected that in the future, in addition to knee and hip prostheses, prostheses of the hand and foot will be introduced. Hopes are especially high with respect to intelligent prostheses. These self-operating prostheses are easy to use, since, by means of biofeedback systems, they function independently of the user.
This brings us to the problem of prostheses for the elderly. One prerequisite for the application of these prostheses is a well-functioning central nervous system. As has already been pointed out, multimorbidity among the elderly is a frequently encountered phenomenon, often accompanied by greatly reduced mental capacities. Multimorbidity implies that precisely the elderly will be able to profit less from this new prosthesis technique. Another problem still to be solved is connecting intelligent prostheses (especially the artificial eye and artificial ear) to the correct nerves. In the case of younger patients this is much less of a problem since the plasticity of the brain is may times greater than in the elderly. Brain functions in the young can thus be taken over much more easily by other parts of the brain. A possible solution to this problem is the presence of various 'reserve receptors' distributed through the body, the use of which is up to now not understood. Instead of the defective receptors, it is usually possible to make use of these reserve receptors. (One example mentioned was 'looking with the stomach', in which the artificial eye is connected to receptors in the stomach which can distinguish between light and darkness).
In addition to prostheses, aids will also be developed which will assist the less mobile elderly individual in the performance of daily tasks (robotics), so that it will be possible for him to function longer and more independently in his own environment. Possibilities also exist for adapted vehicles which, if produced in large numbers, could be kept at a reasonable price.

c. Aids for implantation

High expectations are entertained with respect to aids for implantation. Possibilities which this encompases are drug reservoirs, which once swallowed or implanted (insulin pump) will dispense uniform doses of the drug and function for several days. In addition to this, over a longer period of time it is expected that a solution will be found for the fuelling of the artificial heart as a result of which it will be possible to apply this method on a wider scale (in addition to heart transplants). The electronic heart regulator (pacemaker) is still in its infancy. Here too, it is expected that intelligent apparatus will be developed which is much

smaller and operates better. It seems not unlikely that pacemakers of an electronically high level could become much cheaper through mass production.
In the case of incontinence, artificial sphincter muscles and alarm systems are already implanted. Here too, however, it is very important that the neurologic system of the patient should be in good working order. In addition to the above, there will be a vast increase in the use of functional polymeres, especially organic-chemical substances in transplantation and growth stimulation (for instance prostheses of the jaw and reservoir pills).

d. Medication

Greater insight is expected into the toxicology of drugs among the elderly. Especially insight into the interaction between various drugs, dosage in relation to the age and condition of the patient, monitoring and improvement of administration with the aid of reservoir pills, could result in more effective therapy.

e. Microelectronics

Finally, microelectronics are expected to play an important role in the collection and analyzing of large quantities of medical information so that better insight can be obtained into the prevalence of certain diseases and the possible causes (screening, early diagnosis, etc.).

V CONSEQUENCES FOR THE FUTURE DEVELOPMENT OF HEALTH CARE FACILITIES

General

Making forecasts as to the future range of health care facilities proved to be a difficult task for the participants in the consultation groups. Though the experts often entertained opinions as to the future development of a specific disease, they remained rather vague as to expected future prevalence and the resulting implications for the system of care. For this reason in this section we can only assemble some fragmentary remarks made by the various participants in the course of the discussions.

a. Intramural facilities

A considerable increase in the number of patients is expected in some intramural facilities. This increase can be attributed solely to the future demographic structure of the population. Especially a sharp increase is expected in the number of demented patients. It seems likely that there will be a decrease in the number of people in homes for the elderly as a consequence of the present desire of the elderly to remain as long as possible in their own residential environment.
Thanks to better possibilities for treatment it is expected that there will be an increase in the number of admissions to hospitals. No opinions were expressed as to numbers or percentages and cost developments.

b. Reciprocal aid and umbrella care

Participants were rather negative with respect to the development of this type of aid in the future. Attention was drawn to a survey among the elderly in which there were clear indications that the elderly show a preference for paid help rather than aid from volunteers. The reasons put forward were (assumed) incompetence on the part of the volunteers and the gratitude aspect. Neither were participants very sanguine as to the reappearance of the extended family, in view of the social attitude that the various generations desire to remain independent of each other for as long as possible.

c. Number of medical treatments

As a consequence of better diagnosis methods (inter alia scanning apparatus) and better possibilities for treatment also at an increasingly older age, there will be a large increase in the number of medical treatments performed. This increase is especially observable in the short term. In how far this will lead to an improvement in the health situation over a longer period of time, and thus to a decrease in the number of treatments is a question which is difficult to answer.

d. Mass screening

(See the item malignant neoplasms in the sector on specific diseases.)

e. Influence of developments in other countries on facilities in the Netherlands

It has been repeatedly argued that developments in other countries do not put pressure on Dutch facilities to change (for instance because certain treatments are possible in other countries which cannot be performed in the Netherlands). The level of medical expertise in the Netherlands is such that every known treatment can be offered. The only pressure on the system might derive from the capacity problem, namely that the number of patients for treatment exceeds the admission capacity. This results in waiting lists and criteria for treatment. As far as the participants were aware, up to the present this had resulted in no problems.

PARTICIPANTS IN CONSULTATIONS

1	Dr. A. Berns,	Biochemist, Biochemical Laboratory of the Catholic University of Nijmegen.
2	Dr. C.F. van Bezooyen,	Biologist, Institute of Experimental Gerontology in Rijswijk, Secretary of the Dutch Gerontological Society.
3	Prof. J.C. Birkenhäger,	Professor of Internal Medicine specializing in Metabolic Diseases and Endocrinology, Dijkzigt Hospital, Rotterdam.
4	Dr. A. Bruschke,	Cardiologist, Head of the Department for Cardiology of the Antonius Hospital in Nieuwegein, Chairman of the Dutch Cardiological Society.
5	Prof. S.A. Duursma,	Professor of Internal Medicine specializing in bone metabolism, University Hospital of Utrecht.
6	Dr. E. Fliers,	Physician, Dutch Institute for Brain Research, Amsterdam.
7	Dr. J.A. de Fockert,	Internist, St. Joannes de Deo Hospital, Haarlem.
8	Prof. W.H. Gispen,	Professor of Molecular Neurobiology, University of Utrecht.
9	Dr. J. Haayman	Biologist, Institute of Experimental Gerontology, Rijswijk.
10	Dr. S. Heringa,	Psychiatrist, specialized in psychiatry of the elderly.

11	Dr. H.A. Holtkamp,	Cardiologist, Sophia Hospital, Zwolle.
12	Dr. F. van 't Hooft,	Psychiatrist, Chairman of the Section Psychiatry of the Elderly of the Dutch Society for Psychiatry.
13	Prof. W. Hijmans,	Professor of Immunology, Interuniversity Medical Centre, Leiden.
14	Dr. J. Jolles,	Psychologist and Biochemist, Psychiatric Clinic of the University Hospital, Utrecht.
15	Dr. T. Keuzekamp,	Director of the Nursing Home Schiehoeven, Rotterdam.
16	Dr. W. Kolsters	Cardiologist, Elizabeth Hospital, Amersfoort.
17	Prof. D.L. Knook,	Biochemist and Deputy Director of the Institute of Experimental Gerontology, Rijswijk.
18	Mr. K.J. Kraan, Eng.	Superintendent of Apparatus, Hospital of the Free University, Amsterdam.
19	Dr. C. Leering	Physician in Velp.
20	Mr. P. Lever, Eng.	Superintendent of Apparatus, University Hospital, Utrecht.
21	Dr. H. Mattie,	Internist, Clinical Pharmacologist Department for Infectious Diseases, University of Leiden.

22	Prof. J. Noordhoek,	Professor of Pharmacology, in particular Pharmacokinetics, University of Utrecht.
23	Dr. A. van der Poel,	Biologist and Behavioural Pharmacologist Pharmacological Laboratory, Leiden.
24	Prof. L.B.A. van de Putte	Professor of Internal Medicine, with special emphasis on rheumatology, Catholic University, Nijmegen.
25	Dr. S. Rep,	Urologist, Medical Centre, Alkmaar.
26	Dr. M. Sluyser,	Endocrinologist, Head of the Section Tumor Biology of the Antoni van Leeuwenhoek Centre, Amsterdam.
27	Prof. T.H. The,	Professor of Internal Medicine with special emphasis on clinical immunology, University Hospital, Groningen.
28	Prof. S.L. Visser,	Professor of Clinical Neurophysiology, Hospital of the Free University, Amsterdam.
29	Prof. J.H.P. Wilson,	Professor of Internal Medicine, Dijkzigt Hospital, Rotterdam.
30	Dr. R.J. Zonneveld,	Physician in Oegstgeest, specialized in gerontological research.

APPENDIX J: ESTIMATE OF THE NUMBER OF DEMENTING PERSONS OF 65 AND OLDER IN 1983

Application of the prevalence figures from Table 6.1 to the population structure of 1983* yields the following results.

Age category	Men	
65 - 69	3.9% x 244,754 =	9,545
70 - 74	4.1% x 191,863 =	7,866
75 - 79	8.0% x 130,467 =	10,437
80 +	13.2% x 119,792 =	15,813
	absolute	43,661
	relative	6.36%

Age category	Women	
65 - 69	0.5% x 299,655 =	1,498
70 - 74	2.7% x 266,369 =	7,193
75 - 79	7.9% x 210,342 =	16,617
80 +	20.9% x 228,664 =	47,791
	absolute	73,099
	relative	7.27%

	Total men and women	
65 +	absolute	116,700
	relative	6.90%

---

* Provisional figures CBS

APPENDIX K: ESTIMATE OF THE NUMBER OF PATIENTS WITH THE PRIMARY DIAGNOSIS DEMENTIA IN NURSING HOMES WITH OR WITHOUT A POSTPONEMENT OF DEMENTIA OF FIVE YEARS

In the estimates, use is made of the formula IxD=P (incidence by duration of disease equals prevalence) known in epidemiological literature. The formula assumes an equilibrium situation in the system and is normally applied to **disease** in a population. In this case, the formula is applied to the **nursing home sector**. The content of the concepts however does not change.
Incidence continues to be a **period** variable (that is number of (re)admissions of patients with the primary diagnosis dementia in the nursing home per annum). Prevalence continues to be a **moment** variable (that is, number of patients with the primary diagnosis dementia, in nursing homes at a certain moment).
Lastly, D relates to the **duration** of the process (that is, duration of nursing).

In this appendix the calculation method is given only in outline:

1) In the first place, SIVIS (Nursing Home Information System) data for the years 1981, 1982 and 1983 have been averaged so as to obtain more stable data.

2) Next, number of days in the nursing home is calculated according to age and sex. Though the Nursing Home Information System does provide data as to the number of days spent in the home, these are linked to (the age of) discharge. Since it is preferable to proceed from the duration of stay on the basis of admission (age), these figures must be further adapted. For this purpose use is made inter alia of the admission/discharge matrix. We will not go into the details of this calculation.

3) It is next calculated what would be the consequences for the **present** situation (average of the period 1981-1983) if there should be an immediate five-year postponement of dementia. To this end, incidence figures have been shifted forward by one five-year cohort (the incidence in the age category 65-69 is assumed to be equal to that of the age category 60-64 etc.). The figures for duration of stay have **not** been shifted forward, partly because the duration of stay is probably to a large extent dependent on the age on admission.

If incidence is multiplied by duration of stay, prevalence in a situation of postponement is obtained. By comparison with the situation without postponement, this yields a reduction in dementia of 26%. The calculation is shown below.

I*xD=P
(Incidence figures shifted forward) by (expected duration of sojourn on admission) = (prevalence)

age	men	women
65-69	10 x 3.4 = 34	11 x 5.0 = 55
70-74	51 x 3.0 = 153	64 x 4.8 = 307
75-79	178 x 2.4 = 427	238 x 4.6 = 1095
80-84	349 x 1.8 = 628	622 x 3.7 = 2301
85-89	415 x 1.3 = 540	830 x 2.4 = 1992
90-94	305 x 0.9 = 275	663 x 1.1 = 729
95-99	119 x 0.5 = 60	259 x 0.8 = 207
100 +	18 x 0.1 = 2	44 x 0.5 = 22
Total	2119	6708

Men : Total 2119 with postponement and 2699 without postponement (a reduction of 21%)

Women : Total 6708 with postponement and 9279 without postponement (a reduction of 28%)

Men + women : Total 8827 with postponement and 11978 without postponement (a reduction of 26%).

4) To show what the postponement of dementia will mean in the year 2000 the results of 3) per age and sex category, have been 'inflated' with a factor which corresponds with the increase of the population category between now and 2000 (based on the medium variant of the CBS forecast).

These factors are respectively:

Age	men	women
65-69	1.21	1.12
70-74	1.21	1.16
75-79	1.31	1.30
80-84	1.25	1.31
85 +	1.42	1.82

The reader is referred to Table 6.4 in the main text for the results.

APPENDIX L: ESTIMATE OF THE NUMBER OF INDEPENDENTLY RESIDING ELDERLY PERSONS WHO RECEIVE AID FROM CHILDREN LIVING AWAY FROM HOME IN 1990 AND 2000

Formula: a x (100% - b + c) x d

where:

a = percentage of independently residing elderly who receive aid from children living away from home (see Table 6.5).

b = percentage of institutionalized elderly persons in homes (see also Table 2.17). No further specification according to age category is known for elderly persons under the age of 65 in homes. It is assumed that these all fall within the age category 55-64.

c = percentage of elderly persons institutionalized in nursing homes (based on data of the Nursing Home Information System).

d = absolute number of elderly to be expected according to the medium variant of the CBS forecast.

Men - 1990 -

55-64	3.9% x (100% - 0.1% + 0.2%) x 680,000 =	26,440
65-69	7.2% x (100% - 0.6% + 0.4%) x 282,000 =	20,101
70-74	12.3% x (100% - 2.1% + 1.0%) x 196,000 =	23,361
75-79	24.3% x (100% - 6.5% + 2.0%) x 144,000 =	32,018
80 and older	22.0% x (100% - 21.7% + 5.0%) x 138,000 =	22,254 +

Total men 1990		124,174

Women - 1990 -

55-64	6.1% x (100% - 0.2% + 0.2%) x 717,000 =	43,562
65-69	13.2% x (100% - 0.9% + 0.6%) x 344,000 =	44,727
70-74	13.2% x (100% - 3.5% + 1.3%) x 268,000 =	33,678
75-79	23.8% x (100% - 10.4% + 3.0%) x 232,000 =	47,817
80 and older	25.9% x (100% - 29.7% + 8.5%) x 291,000 =	46,578 +

Total women 1990		216,362

Total men and women 1990 : 340,536.

Men - 2000 -

55-64	3.9% x (100% - 0.1% + 0.2%) x 789,000 =	30,679	
65-69	7.2% x (100% - 0.6% + 0.4%) x 296,000 =	21,099	
70-74	12.3% x (100% - 2.1% + 1.0%) x 230,000 =	27,413	
75-79	24.3% x (100% - 6.5% + 2.0%) x 169,000 =	37,576	
80 and older	22.0% x (100% - 21.7% + 5.0%) x 155,000 =	24,995	+

Total men 2000		141,762	

Women - 2000 -

55-64	6.1% x (100% - 0.2% + 0.2%) x 785,000 =	47,693	
65-69	13.2% x (100% - 0.9% + 0.6%) x 336,000 =	43,687	
70-74	13.2% x (100% - 3.5% + 1.3%) x 304,000 =	38,202	
75-79	23.8% x (100% - 10.4% + 3.0%) x 268,000 =	55,237	
80 and older	25.9% x (100% - 29.7% + 8.5%) x 331,000 =	52,981	+

Total women 2000		237,800	

Total men and women 2000 : 379,562

APPENDIX M: FLOW OF INDEPENDENTLY RESIDING ELDERLY PERSONS TO FACILITIES IF AID FROM CHILDREN LIVING AWAY FROM HOME COMPLETELY FALLS OFF IN 1990 OR 2000

1990

67%	x 340,000 :	50% to first echelon facilities	:	113,900
		10% to adapted dwellings	:	22,780
		40% to commercial aid	:	91,120
18.9%	x 340,000 :	50% to homes for the elderly	:	32,130
		40% to first echelon facilities	:	25,704
		10% to commercial aid	:	6,426
14.1%	x 340,000 :	50% to first echelon facilities	:	23,970
		50% to commercial aid	:	23,970

2000

67%	x 380,000 :	50% to first echelon facilities	:	127,300
		10% to adapted dwellings	:	25,460
		40% to commercial aid	:	101,840
18.9%	x 380,000 :	50% to homes for the elderly	:	35,910
		40% to first echelon facilities	:	28,728
		10% to commercial aid	:	7,182
14.1%	x 380,000 :	50% to first echelon facilities	:	26,790
		50% to commercial aid	:	26,790

Glossary

A

ADL	Activities of Daily Life.
Ageing	Increase of the share of elderly (55-plus or 65-plus) in the total population of a country; cf. double ageing.

C

Cohort	All individuals born in the same year (birth cohort) or married in the same year (marriage cohort).
Context scenario	Indicates the autonomous developments within which an actor must weigh alternatives for action and must ultimately implement his plans (scenario in a narrow sense).

D

Demographic pressure

$$\frac{\text{0-19-year-olds + 65-plussers}}{\text{20-64-year-olds}} \times 100\%$$

The age limit of 65 can also be replaced by 55.

Double ageing

Increase in the share of the elderly in the total population, plus an increase of the share of the very aged (80-plus) in the category of elderly; compare ageing.

E

Early warning system	(Also called 'look-out institution'), an institution in search of policy issues which are not recognized as such because of unawareness of possible developments and their significance (I.Dror, Ventures in Policy Sciences, New York, 1971).
Epidemiology	The science concerned with the study of the frequency and distribution of diseases in connection with the underlying causes.
Ex ante evaluation	Literally 'judging beforehand': the systematic study beforehand of what effects and side-effects might be expected from the introduction of a certain policy measure.

F

Forecast	A deductive argumentation along strictly scientific lines of the future state of a given situation on the basis of hypothesis and initial conditions (J. van Doorn and F. van Vught, Forecasting, Assen, 1978, p. 14).

I

Incidence	(Percentage of) the number of new cases of a certain disease in a society during a given period.
Institutionalization	The transition from residing independently to living in an institution such as a home for the elderly or nursing home.

L

Lifestyle	Way of life. In the context of this report especially habits involving risk, such as heavy smoking, excessive use of alcohol and too little exercise.

M

Monitoring	Following social developments over longer periods of time with the aid of statistical information (social indicators etc.).
Multiple pathology	The presence of diverse diseases or complaints in one patient.

P

Palliation	Assuaging the suffering brought about by a disease without curing the disease itself.
Pensioning	Flexible: the ability to retire at will between for instance the ages of 60 and 70. Gradual or half-pensioning: (half)pension is supplemented by income from a part-time job.
Policy	A more or less well-considered effort to achieve certain aims with certain tools and in a certain order of sequence (A. Hoogerwerf (ed.), Policy Explained, Part I, Alphen on the Rhine, 1972, p. 84). Note: here policy is taken in a wide sense and includes for instance government and politics.

Pressure of the elderly	$\frac{\text{65-plussers}}{\text{20-64-year-olds}} \times 100\%$ The age limit of 65 can be replaced by 55.
Prevalence	The total number of persons in a population suffering from a certain disease at a given moment.
Prevention	Primary: preventing disease. Secondary: recognizing diseases in an early stage (early diagnosis).
Professionalization	Monopolization by professionals of activities which were formerly performed by volunteers or untrained persons.

Q

Quaternary sector	Non-profit service sector (other sectors are: the tertiary sector: commercial services; secondary sector: industry; primary sector: agriculture and fishery).

R

Reciprocal aid	Aid provided each other by family, friends or neighbours (cf. umbrella care).
Reference scenario	A scenario which serves as a background for other scenarios. - In the case of a null variant, one or more components are assumed to remain permanently or temporarily constant. - In the case of a trend variant, developments from the past are taken into account in future developments.

- In the case of a policy variant, future developments are described which are based on known government policy.

S

Scenario in a narrow sense	Possible future situations in a society which arise more or less autonomously of the society (or part of it) (context scenario).
Scenario in a wider sense	A description of the present state of a society (or part of it), of possible and desirable future states of the society, as well of series of events which could result in the present state evolving into the future state (Department of Planning and Policy, A Manual for the Designing of Scenarios, Utrecht, 1981).

T

Technology assessment	Ex ante evaluation (see above) of the effects of technological innovations (for instance in the sphere of biotechnology).
Telediagnosis	Making of diagnosis by means of telecommunication; for instance feeding the symptoms of a disease to a computer and later receiving information regarding the disease, the desirability of consulting a specialist, etc.

U

Umbrella care	The general eye kept on the elderly and aid provided by relations, friends and neighbours (cf. reciprocal aid).

Bibliography

Achterhuis, H.: The Market of Well-being and Happiness. Baarn, 1979. (in Dutch only)

Adriaansens, H.P.M. and A.C. Zijderveld: Voluntary Initiative and the Welfare State. Deventer, 1981. (in Dutch only)

Bakker-Lenderink, A.: The Elderly in Cross Society Care and the Consequences of Growth and Economization. In: Monthly Bulletin Social Health Care, Vol. 11, No. 9, September 1983, pp. 8-11. (in Dutch only)

Bartels, L.P.: Institutions for Intramural Health Care. Part 1: Basic Data 1.1.1984. Hospitals Institute of the Netherlands, Utrecht, 1984. (in Dutch only)

Beck, P.W.: Forecasts: Opiates for Decision Makers. Paper Third International Symposium on Forecasting. Philadelphia, June 1983.

Berg, P.: Biotechnology: A Challenge for the Future. Jubilee Conference on Education and Science. The Hague, August 1983. (in Dutch only)

Biesenbeek, F.J.: What Becomes of the Nursing Home? The Danish Model. In: Medical Contact, Nos. 1-7, January 1983, pp. 22-24. (in Dutch only)

Boelen, K.W.J. and G.J. Zwanikken: Mental Disorders Among the Elderly, Leiden, 1975. (in Dutch only)

Boruch, R.F. and H.W. Riecken (eds.): Experimental Testing of Public Policy. Boulder, 1975.

Braam, G.P.A., J.A.I. Coolen and J. Naafs: The Elderly in the Netherlands, Sociology of the Elderly, Care of the Elderly and Policy Relating to the Elderly. Alphen on the Rhine/Brussels, 1981. (in Dutch only)

Brouwer, W.: The Single Geriatric Consultation. In: GP and Science 26(1983)4, pp. 140-144. (in Dutch only)

Burg, C.: The Prospects for Medicine in the Next 20 Years. Second European Seminar for Leading Public Health Administrators on Health for All by the Year 2000. Oslo, August 1983.

CBS: The Handicapped, The Hague, 1976. (in Dutch only)

CBS: The Living Conditions of the Dutch Population of 55 and Older - 1976. Parts 1-6. The Hague, 1977, 1978, 1979, 1981 and 1983. (in Dutch only)

CBS: Forecast of the Population of the Netherlands after 1980. Part 1: Results and Some Backgrounds. The Hague, 1982. (in Dutch only)

CBS: Monthly Population Statistics, 1982/5. (in Dutch only)

CBS: Dental Prostheses Among the Dutch Population. In CBS Monthly Bulletin on Health, 82/10, pp. 5-14. (in Dutch only)

CBS: Statistics of Homes for the Elderly 1979 and 1980. The Hague, 1983. (in Dutch only)

CBS: Living Conditions of the Dutch Population of 55 and Older - 1982. Part 1A: Key Figures. The Hague, 1984. (in Dutch only)

CBS: The Living Conditions of the Dutch Population - 1980. Key Figures. The Hague, 1984. (in Dutch only)

CBS: Monthly Bulletin of Health Statistics 1984/10. (in Dutch only)

Claessens, W.L.M.: Multidisciplinary Observation of Suspected Psychogeriatric Patients. Lisse, 1984. (in Dutch only)

Collette, H.J.A. et al.: Evaluation of Screening for Breast Cancer in a Non-randomised Study by Means of a Case Control Study. In: The Lancet, June 2, 1984, pp. 1224-1226.

COSBO: Fundamental Policy with Respect to the Elderly. Utrecht, 1983. (in Dutch only)

DHV Advisory Bureau: Report of Research as to Intensifying Indication Policy for Homes for the Elderly in the Province of Utrecht. Amersfoort, 1984. (in Dutch only)

Dooghe, G. and R. Bruynooghe: Comments on Knipscheer. In: Health and Society, Vol. 5, No. 2, 1984, pp. 90-92. (in Dutch only)

Doorn, J. van and F. van Vught: Forecasting. Assen, 1978. (in Dutch only)

Dror, I.: Ventures in Policy Sciences. New York, 1971.

Dutch Institute of General Practitioners: Co-operation and Referral. Part 1: Form of Practice and Production Figures. Utrecht, 1983. (in Dutch only)

Dutch Foundation for Care of the Elderly (NFB): Nomenclature of Facilities for the Elderly. The Hague, 1974. (in Dutch only)

Duve, C. de: Towards a Second Medical Revolution. Jubilee Conference on Education and Science. The Hague, August, 1983. (in Dutch only)

Earth Year 2050, BBC. N.p. 1984.

Fairweather, G.W. and L.G. Tornatzky: Experimental Methods for Social Policy Research. New York, 1977.

Fliers, E., A. Lisei and D.F. Swaab: Dementia: Current Concepts and Research in the Netherlands. Synopsis of a Review Study commissioned by the Dutch Institute of Medicine, Amsterdam. 1983a. (in Dutch only)

Fliers, E., A. Lisei and D.F. Swaab: Dementia. Some Current Concepts and Research in the Netherlands. Amsterdam, 1983b.

Frinking, G.A.B.: Childlessness in Figures; Trends in Voluntary and Involuntary Childlessness. In: Intermediair 11(1975)46 (14 November). (in Dutch only)

Fuldauer, A.: The GP and Research on the Elderly. In: The GP and Science, 1968, 11, pp. 99-102. (in Dutch only)

Fuldauer, A.: Nine Years of Research on the Elderly in a GP Practice: A Report on Preventative Medical Research. In: GP and Science, 1973, 16, p. 135-147. (in Dutch only)

Fuldauer, A. et al.: Screening of the Elderly. Five Years of Multidisciplinary Psychogeriatric Research. In: Monthly Bulletin on Mental Health, 1980, 3, pp. 210-215. (in Dutch only)

Fuldauer, A. and J. Langendijk: A Study as to the Degree of Mental Handicap Among Residents of Homes for the Elderly, 1980 (unpublished). (in Dutch only)

Genuchten, H.J.M. van: Psychogeriatric Problems in GP Practices in Amsterdam. Department of Social Gerontology, Nijmegen, June 1983. (in Dutch only)

Godderis, J.: Depression and the Ageing. In: Journal of Psychiatry, 25, 1983/5, p. 303-332. (in Dutch only)

Goedhard, W.J.A.: Epilogue. In: Knook, D.L. And W.J.A. Goedhard (eds.): Dementia and Ageing of the Brain. Alphen on the Rhine/ Brussels, 1981, pp. 135-140. (in Dutch only)

Goudriaan, R. et al.: Collective Expenditure and Demographic Developments 1970-2030. S.C.P. Working Paper No. 38. The Hague, 1984. (in Dutch only)

Goudriaan, R., H. de Groot et al.: Trend Report of the Quaternary Sector 1983-1990, S.C.P. Working Paper No. 43, The Hague, 1984. (in Dutch only)

Groot, H. de, H. van der Meer and J.M.M. Ritzen: Social Services and Health Care. S.C.P. Working Paper 24. Rijswijk, 1981. (in Dutch only)

Habbema, J.D.F., et al., No Difference in Results Between Women Under and Over the Age of 50 in a General Screening of the Population for Cancer of the Breast. In: Journal of Social Health Care 61(1983)19, pp. 694-697. (in Dutch only)

Harmonization Council for Welfare Policy: Limits of Health Care. No. 33. The Hague, November 1983. (in Dutch only)

Hattinga Verschure, J.C.M.: The Phenomenon Care. Lochem, 1977. (in Dutch only)

Havighurst, R.J.: Social Change. The Status, Needs and Wants of the Future Elderly. In: B. Nieman Herzog: Aging and Income Programs and Project for the Elderly. New York, 1978.

Heuvel, W.J.A. van den: The Meaning of Dependency. In: J.M.A. Munnichs and W.J.A. van den Heuvel (eds.): Dependency or Interdependency in Old Age. The Hague, 1976, pp. 162-173.

Hilhorst, H.W.A. and M.J. Verhoef: The 'Easy Death' in Practice. Utrecht, 1979. (in Dutch only)

Hollander, C.F.: Biological Research as to Ageing Important for the Elderly Patient. In: TNO Project, Vol. 12, Nr. 5 - June 1984, pp. 233-236. (in Dutch only)

Hoogen, H.J.M. van den et al.: GP and Ageing. In: Journal for Social Medicine, 60(1982), No. 25, pp. 870-874. (in Dutch only)

Hoogendoorn, D.: Some Remarks as to the Significance of Ageing of the Population for Health Care. In: Dutch Journal of Medicine, 121, No. 42, 1977, pp. 1639-1643. (in Dutch only)

Hoogerwerf, A. (ed.): Policy Explained. Part 1. Alphen on the Rhine, 1972. (in Dutch only)

Horn, S. ten, The Need for Psychogeriatric Care. In: Monthly Bulletin for Mental Health, 3/81, pp. 205-214. (in Dutch only)

Hospitals Institute of the Netherlands: Report on a Survey of the Question of 65-Plussers. Internal memorandum of the Hospitals Institute, March 1981. (in Dutch only)

Hospitals Council of the Netherlands: Discussion Memorandum 'Duties and Function of Nursing Homes. An Impulse to Renewal'. Utrecht, July 1981. (in Dutch only)

Houben, P.: Desirable Residential Solutions According to the Elderly, Practical Experts and Policy. In: Senior, No. 20/1983, pp. 488-491. (in Dutch only)

Houben, P.P.J.: Social Developments and Housing for the Elderly. In: J.B. Burie (ed.): Housing and the Elderly. SOOM Working Paper No. 1. Nijmegen, 1984. (in Dutch only)

Houben, P.P.J., F. Wind and H. Moeskops: Residential Needs of the Elderly Examined: New Criteria for Policy. Delft/The Hague, 1984. (in Dutch only)

Inderdepartmental Committee for Volunteer Policy: Volunteer Policy. Reports 1 - 3. The Hage, 1980-1982. (in Dutch only)

Janerich, D.T.: Forecasting Cancer Trends to Optimize Control Strategies. In: Journal of the National Cancer Institute, Vol. 72, No. 6, June 1984, pp. 1317-1321.

Kam, C.A. de and F.P. van Tulder: 'Cross Your Bridges When You Reach Them?' In: The Elderly and Social Security. The Hague, 1983, pp. 33-64. (in Dutch only)

Katz, S. et al.: Active Life Expectancy. In: The New England Journal of Medicine, Nov. 17, 1983, pp. 1218-1224.

Katzman, R.: The Prevalence and Malignancy of Alzheimer Disease - A Major Killer. In: Arch. Neurol. 33, 1976, pp. 217-218.

Katzman, R., R.D. Terry and K.L. Bick (eds.): Alzheimer Disease. Senile Dementia and Related Disorders. Aging Series, Vol. 7. New York, 1978.

Kay, D.W.K. and K. Bergmann: Epidemiology of Mental Disorders Among the Aged in the Community. In: J.E. Birren: Handbook of Mental Health and Aging, 1980, pp. 34-56.

Knipscheer, C.P.M.: Family Care within Policy for the Elderly: The State of Affairs and Developments. In: Health and Society, Vol. 5, No. 2, 1984, pp. 80-89. (in Dutch only)

Knook, D.L.: Research as to Ageing in Europe: Ten Heads are Better Than One. In: TNO Project, Vol. 12, No. 5, June 1984, pp. 244-246. (in Dutch only)

Kronjee, G.J.: Normative Points of Departure for a Research Programme as to the Residential Situation of the Elderly. In: J.B. Burie (ed.): Housing and the Elderly, SOOM Working Paper No. 1. Nijmegen, May 1984. (in Dutch only)

Lägergren, M., et al.: Time to Care. Oxford, 1984.

Langeveld, H.M.: Ageing and Work. Programming Advice. SOOM Working Paper No. 2, Nijmegen, April 1984. (in Dutch only)

Luteijn, F., J. Niemeyer, D.H. Sipsma, H.W. ter Haar: A Differentiation of the Elderly According to Degree of Mental Derangement. In: Dutch Journal of Gerontology, 1972, pp. 314-326. (in Dutch only)

Luyk, E.W. van and R.J. de Bruijn: Volunteer Work Between Paid and Domestic Work: An Exploratory Study on the Basis of a Survey. Scientific Council for Government Policy, The Hague, 1984. (in Dutch only)

Maas, P. van der: Myths as to Ageing and Health. In: Journal of Social Medicine, 60(1982), No. 23, pp. 711-721. (in Dutch only)

Makinodan, T.: Biology of Aging: Retrospect and Prospect. In: T. Makinodan and E. Yunis (eds.): Comprehensive Immunology and Aging. New York, 1977, pp. 1-7.

Meer. D. van der: Policy and Research With Respect to Intramural Health Care. May 1983. (in Dutch only)

Memorandum of First Echelon Care. Second Chamber, 1983/1984 Session, 18180, Nos. 1 and 2. (in Dutch only)

Miesen, B.: Outpatient Somatopsychosocial Screening of Behavioural Disturbances among the Aged. In: Dutch Journal of Gerontology, 5; 3, 1974, pp. 153-158. (in Dutch only)

Minister of Culture, Recreation and Social Work, et al.: Aspects of Policy Relating to the Elderly. The Hague, 1982. (in Dutch only)

Minister of Welfare, Health and Culture, et al., Note on Policy Relating to the Elderly. Leidschendam, 1983. (in Dutch only)

Ministry of Internal Affairs/SCP: Final Report of the Committee for the Harmonization of Estimates for the Quaternary Sector plus Technical Report. The Hague, September 1983. (in Dutch only)

Minister of Welfare, Health and Culture: Flanking Policy for the Elderly. Leidschendam, 1983. (in Dutch only)

Mootz, M. and J. Timmermans: Taking Care of the Future. Desiderata for a Future Policy for the Elderly. SCP Working Paper No. 26. Rijswijk, 1981. (in Dutch only)

Murtomaa, M. and J. Kankaanpää: Scenarios for the HFA 2000. Helsinki, 1984.

National Association of Cross Organizations: Multy-year Estimate of Work of Cross Societies 1985-1988. Bunnik, February 1984 (in Dutch only)

Niphuis-Nell, M.: Characteristics of Voluntarily Childless Women in the Netherlands. In: Population and Family, 1979, 2, pp. 201-205. (in Dutch only)

Nuy, M.H.R., et al.: Day Treatment in Nursing Homes, Parts 1 to 3, Nijmegen, 1984. (in Dutch only)

NVAGG/GHIGV (Dutch Society for Ambulatory Mental Health Care/Chief Medical Inspectorate of Mental Health): The Socio-geriatric Function Within Ambulatory Mental Health Care. N.p., 1979. (in Dutch only)

Pannenborg, C.: Scenarios as a Method of Probing and Planning the Future of Health Care. Paper presented at the WHO conference on Planning and Management for Health, The Hague, 1984.

Pommer, E. and C. Wiebrens: Costs and Financing of Facilities for the Elderly 1981-1991. SCP Working Paper No. 39. The Hague, 1984. (in Dutch only)

Research Group Planning and Policy-making: Research on the Scenario Method. Final Report 1. Manual for the Designing of Scenarios. Utrecht, 1981.

Rigter, H.: Research as to Ageing of the Memory among Animals. In: D.L. Knook and W.J.A. Goedhard (eds.): Dementia and Ageing of the Brain. Alphen on the Rhine/Brussels, 1981, pp. 122-134. (in Dutch only)

Ringoir, D.J.B.: Epidemiological Aspects of Senile Dementia. In: D.L. Knook and W.J.A. Goedhard (eds.): Dementia and Ageing of the Brain, Alphen on the Rhine/Brussels, 1981. (in Dutch only)

Ringoir, D.J.B. and R. van Duuren: An Estimate of the Costs of Health Care for Dementing Patients. In: The Netherlands Journal of Gerontology, 12, 1981, 1, pp. 28-37. (in Dutch only)

Robertson, J.: Scenarios for Lifestyles and Health. N.p., January 1983.

Rongen, M.J.T. and J.M.J.F. Houben: The Housing Situation of the Elderly, The Hague, 1984. (in Dutch only)

Santvoort, M.M. van: Research as to Care Facilities for the Elderly. A Review. SOOM Working Paper No. 3, Nijmegen, 1984. (in Dutch only)

Scenario Committee on Coronary and Arterial Diseases: Summary of Some Cardiovascular Disease Scenarios. The Hague, 1984.

Scenario Committee on Oncology: Summary of Some Cancer Scenarios. The Hague, 1984.

Scenario Committee of Lifestyles: Summary of Some Scenarios of Lifestyles and Health. The Hague, 1984.

Scenario Committee on Ageing: The Elderly and Their Health in The Netherlands, 1984-2000. Towards a Scenario Report. Utrecht, 1984.

Schnabel, P.: New Relationships Between Citizen and State. In: P.A. Idenburg (ed.): The Last Days of the Welfare State. Amsterdam/ Rijswijk, 1983, pp. 25-67. (in Dutch only)

Schouten, J.: The Results of the Department for Somatic and Psychogeriatric Geriatrics of the Slotervaart Hospital, Amsterdam. In: Dutch Medical Journal, 123; 16, 1979, pp. 653-657. (in Dutch only)

Scientific Journal for Government Policy: Re-evaluation of Welfare Policy. The Hague, 1982. (in Dutch only)

SCP: Social and Cultural Report 1984. The Hague, 1984. (in Dutch only)

Secretary of State of Welfare, Health and Culture: Health Policy with Limited Resources. The Hague, 1983. (in Dutch only)

Serail, S.: A Real Home for the Elderly. Tilburg, 1982. (in Dutch only)

Sillevis Smit, W.G.: Stenosed Old Age. Leiden, 1975. (in Dutch only)

Society of Actuaries: Flexible Pensioning. A Study of the Financial Consequences in the Netherlands. Amsterdam, June 1983. (in Dutch only)

Swedisch Ministry of Health and Social Affairs: Health in Sweden. Facts from Basic Studies under the HS 90 Program. Stockholm, 1982.

United Nations: European Social Development Program, Informal Action for the Welfare of the Aged. Seminar Copenhagen, 26 March to 3 april 1979. New York, 1980.

Verbeek, A.L.M. et al.: Reduction of Breast Cancer Mortality Through Mass Screening With Modern Mammography. In: The Lancet, June 2, 1984, pp. 1222-1224.

Voorn, T.: Chronic Diseases in a GP Practice. Utrecht, 1983. (in Dutch only)

Vught, F. van: Experimental Policy Planning. The Hague, 1982. (in Dutch only)

Weeda, C.J.: Dynamics in Household Types. In: P.A. Idenburg (ed.): The Last Days of the Welfare State. Amsterdam/Rijswijk, 1983. (in Dutch only)

Werff, A. van der: Planning and Management for Health in Periods of Economic Stringency and Instability. The Hague, 1984.

Wersch-van der Spek, M.C.M. van: A Geriatric Outpatient Department. In: Medical Contact, 35; 38, 1980, pp. 1161-1167. (in Dutch only)

WHO: Epidemiological Studies on Social and Medical Conditions of the Elderly. Copenhagen, 1982a.

WHO: Preventing Disability in the Elderly. Copenhage, 1982b.

WHO: Epidemiology of Aging. Geneva, September 1983 (Draft).

Willemse, P.M.A.: Geriatric Care in the GP Practice. The Single Consultation. In: Medical Contact 36(1981)22, pp. 667-669. (in Dutch only)